T0310229

Electromagnetic Time Reversal

Electromagnetic Time Reversal

Application to Electromagnetic Compatibility
and Power Systems

Edited by

Farhad Rachidi
Swiss Federal Institute of Technology (EPFL), Lausanne, Switzerland

Marcos Rubinstein
University of Applied Sciences of Western Switzerland, Yverdon,
Switzerland

Mario Paolone
Swiss Federal Institute of Technology (EPFL), Lausanne, Switzerland

Registered Offices
John Wiley & Sons, Inc., 111 River Street, Hoboken, NJ 07030, USA
John Wiley & Sons Ltd, The Atrium, Southern Gate, Chichester, West Sussex, PO19 8SQ, UK

Editorial Office
The Atrium, Southern Gate, Chichester, West Sussex, PO19 8SQ, UK

For details of our global editorial offices, customer services, and more information about Wiley products visit us at www.wiley.com.

Wiley also publishes its books in a variety of electronic formats and by print-on-demand. Some content that appears in standard print versions of this book may not be available in other formats.

Library of Congress Cataloging-in-Publication Data

Names: Rachidi, Farhad, 1962– editor. | Rubinstein, Marcos, 1960– editor. | Paolone, Mario, 1973– editor.
Title: Electromagnetic time reversal : application to electromagnetic compatibility and power systems / edited by, Farhad Rachidi, Marcos Rubinstein, Mario Paolone.
Description: Hoboken, New Jersey : John Wiley & Sons, Inc., [2017] | Includes bibliographical references and index.
Identifiers: LCCN 2016042856 (print) | LCCN 2016052640 (ebook) | ISBN 9781119142089 (cloth ; alk. paper) | ISBN 1119142083 (cloth ; alk. paper) | ISBN 9781119142096 (Adobe PDF) | ISBN 9781119142102 (ePub)
Subjects: LCSH: Time reversal. | Electromagnetism. | Electromagnetic compatibility. | Electric power systems.
Classification: LCC QC173.59.T53 E44 2017 (print) | LCC QC173.59.T53 (ebook) | DDC 621.3101/53–dc23
LC record available at https://lccn.loc.gov/2016042856

Cover image: Kaisorn/Gettyimages
　　　　　　Sylverarts/Gettyimages

Set in 10/12pt WarnockPro by Aptara Inc., New Delhi, India
Printed and bound in Malaysia by Vivar Printing Sdn Bhd

10 9 8 7 6 5 4 3 2 1

"Lost time is never found again" Benjamin Franklin

Contents

List of Contributors

Pierre Bonnet
Blaise Pascal University

Andrea Cozza
Group of Electrical
Engineering Paris (GeePs)

Matthieu Davy
University of Rennes

Julien de Rosny
Langevin Institute

Mathias Fink
Langevin Institute

Sébastien Lalléchère
Blaise Pascal University

Gaspard Lugrin
Swiss Federal Institute of
Technology of Lausanne
(EPFL)

Florian Monsef
Group of electrical
engineering Paris (GeePs)

Michel Ney
Telecom Bretagne

Pascal Paganini
Telecom Bretagne

Françoise Paladian
Blaise Pascal University

Mario Paolone
Swiss Federal Institute of
Technology of Lausanne
(EPFL)

Farhad Rachidi
Swiss Federal Institute of
Technology of Lausanne
(EPFL)

Reza Razzaghi
Swiss Federal Institute of
Technology of Lausanne
(EPFL)

Marcos Rubinstein
University of Applied
Sciences of Western
Switzerland

Ahmed Zeddam
Orange Labs

Preface

Time reversal has emerged as a very interesting technique with potential applications in various fields of engineering. It has received a great deal of attention in recent years, essentially in the field of acoustics, where it was first developed by Prof. Fink and his team in the 1990s. In the past decade, the technique has also been used in the field of electromagnetics and applied to various other areas of electrical engineering. In particular, the technique has been successfully applied in the fields of electromagnetic compatibility (EMC) and power systems, leading to mature technologies in source-location identification with unprecedented performance compared to classical approaches. It is expected that the fields of application of electromagnetic time reversal (EMTR) will continue to grow in the near future.

This book is intended to give the theoretical foundation of the electromagnetic time-reversal theory. Special emphasis is given on real applications in the fields of EMC and power systems.

The book's introductory chapter presents the theoretical basis of the electromagnetic time-reversal technique. It starts with a discussion of the notion of time in physics and goes on to present three approaches that can be used to effectively make a system *go back in time*, in the sense that it retraces the path it came from in the immediate past. The concepts of strict and soft time-reversal invariance are introduced and illustrated using simple examples. The time-reversal invariance of physics laws is then described, with special attention given to the time-reversal invariance of Maxwell's equations. The concept of time-reversal cavity and the use of time reversal as a means of refocusing electromagnetic waves is then described. The chapter ends with a brief presentation of application areas of electromagnetic time reversal.

Chapters 2 to 7 are devoted to specific applications of EMTR, including EMC measurements, EM field focusing and amplification, interference mitigation in power line communications, lightning detection, and fault location in power systems.

In Chapter 2, the potential use of time reversal in diffusive media for radiative testing is addressed, in particular for EMC, antenna testing, and channel emulation. The chapter starts with a brief review of common features of diffusive media, introducing probabilistic models for the random nature of fields in them, showing the complexity of the media and making the case for the generation of coherent wavefronts. The response of a diffusive medium to time-reversed signals is then analyzed for point-to-point transmissions, illustrating how received signals are affected by background random fluctuations due to the frequency-selective response of the medium. It is shown that narrow bandwidths are sufficient to enable the properties of time reversal, a point that is of fundamental importance in high-power microwave applications. Other properties are then presented, including the possibility of using single-antenna time-reversal (TR) mirrors and the ability of TR to control the polarization of received fields, independent of the features of the transmitting antenna. In addition, spatial and time focusing are shown to lead to energy efficiencies even higher than those expected in reverberation chambers. Virtual sources are introduced based on the observation that standard TR assumes the availability of sources of radiation whose fields will then be time-reversed. In the final part of the chapter, a generalization of TR is presented that allows the generation of complex, arbitrary wavefronts.

Chapter 3 deals with the robustness of the EMTR process for EMC applications. A number of studies have been carried out covering a broad range of domains, including communications, imaging, and field enhancement. In this framework, electromagnetic compatibility (EMC) may stand to benefit greatly from EMTR, since this technique allows a heretofore-unachievable level of control of electromagnetic waves. This could reduce time and costs during EMC standard tests, assuming that external conditions (antenna location, environment, devices under test) are perfectly known. Few studies have dealt with the potential impact of randomness on EMTR. In this chapter, the main emphasis will be on the accuracy and robustness of using EMTR in practical experiments dealing with immunity, controlling EM fields, and transmission lines.

Traditional focusing systems of wideband signals make use of a beamforming method applied to an array of antennas. In Chapter 4, a different approach, based on the time-reversal technique, is presented to focus high-amplitude wideband pulses. The time-reversal process consists of two phases. First, the transient response between a source outside a cavity and an array of antennas within the cavity is measured. For wideband pulses, the signals spread over a time much longer than the initial pulse length because of the reverberation within the cavity. The signals are then flipped in time and re-emitted. Due to the reversibility of the wave in the propagation medium, the time-reversed field focuses both in time and space at the initial source position. The gain in amplitude of the focused signal is linked to the time compression of the transient response and can therefore be several orders of magnitude higher than the amplitude generated using a beamforming method without the chamber. An analysis of the properties of the focal spot with respect to the different experimental parameters, such as the number of antennas, the aperture, and the size of the cavity or the source polarization is presented. The one-bit time-reversal method to enhance the amplitude of the focused signal is also described. Finally, we show an extension of the method to focus a pulse at any position outside a cavity from the knowledge of the transient responses only in the aperture area.

Chapter 5 presents the use of the TR technique to mitigate radiated emissions from power line communication (PLC) systems. Power line communication is an effective response to today's high demand for multimedia services in the domestic environment, not only for its fast and reliable transfer characteristics but also for its flexible, low-cost implementation, since the PLC technology uses the existing electrical network infrastructure and the ubiquitous outlets throughout the home. In current PLC systems, the high bit rate transfer through the mains network generates acceptable radiated emissions regulated by international standards, but the demand for greater speeds in new generation PLC systems may cause higher levels of emissions. The way in which this method has been experimentally verified in real electrical networks is presented. The second part of this chapter presents the level of effectiveness of TR in reducing the average electromagnetic nterference (EMI) generated by PLC transmissions by combining the effects of channel gain and spatial filtering.

Chapter 6 is devoted to lightning location using EMTR. The first part of this chapter presents a brief overview of the main classical lightning location techniques. Next, the lightning location by the EMTR method is described, followed by a mathematical proof and simulation-based verifications. Then the important issue of the application of EMTR in the presence of losses due to propagation over a finitely conducting ground is dealt with. The relation between EMTR and the difference in time-of-rrival technique is also presented. The last part of the chapter is dedicated to practical implementation issues.

In Chapter 7, we present the use of the EMTR theory for locating faults in both transmission and distribution power networks characterized by meshed and radial topologies. The fault location functionality is an important online process required by power systems operation since, for the case of transmission grids, it has a large influence on the security and, for distribution systems, on the quality of supply. Compared to other transient-based fault location techniques, the EMTR method presents a number of advantages, namely, its straightforward applicability to inhomogeneous media (mixed overhead and coaxial power cable lines), the use of a single observation (measurement) point, and robustness against fault type and fault impedance. All these aspects are presented and discussed in the chapter via simulations and experimental validations of the EMTR-based fault location process.

To the best of our knowledge, this is the first book giving an overview of the EMTR technique and its engineering applications to power systems and EMC. Within the context of the evolution of power networks towards smart grids and the importance of the security and reliability of future grids, we are convinced that EMTR-based techniques described in the book and possibly others developed in the future will find an ever growing field of application.

The editors are indebted to many individuals for their support, advice, and guidance. Special thanks are due to Steven Anlage, Giulio Antonini, Gerhard Diendorfer, Jean Mahseredjian, Hamid Karami, Carlo Alberto Nucci, Antonio Orlandi, Wolfgang Schulz, Keyhan Sheshyekani, Mirjana Stojilovic, Felix Vega, and Yan-Zhao Xie, and to all the authors of the chapters for their precious contributions. Thanks are also due to Asia Codino, Gaspard Lugrin, Hossein M. Manesh, Razieh Moghimi, Nicolas Mora, Andrea Pollini, Reza Razzaghi, and Zhaoyang Wang, who have worked on various aspects of electromagnetic time reversal during their graduate studies, and to undergraduate students Amir Fouladvand and Dara Sadeghi.

Farhad Rachidi, Marcos Rubinstein, and Mario Paolone

About the Companion Website

Electromagnetic Time Reversal: Application to EMC and Power Systems is accompanied by a companion website:

www.wiley.com/go/rachidi55

The website includes:

- Supplementary video animations.

1

Time Reversal: A Different Perspective

M. Rubinstein,[1] F. Rachidi,[2] and M. Paolone[2]

[1] *University of Applied Sciences of Western Switzerland, Yverdon, Switzerland*
[2] *Swiss Federal Institute of Technology (EPFL), Lausanne, Switzerland*

1.1 Introduction

In this introductory chapter, we will present the theoretical background of the time reversal in electromagnetism. We will start by discussing the notion of time in physics. The time-reversal invariance of physics laws will then be described. Special attention will be devoted to the time-reversal invariance of Maxwell's equations. The concept of time-reversal cavity and the use of time reversal as a means of refocusing electromagnetic waves will be described. Finally, the chapter will end by briefly presenting application areas of electromagnetic time reversal.

1.2 What is Time?

Most of the words that define very complex concepts are rarely used in our daily language and writings.[1] There are, however, exceptions, *time* being a particularly notable one. Even though *time* appears to be a quite familiar notion that carries the feeling

---

1 For instance, the word "inchoate" (partly in existence or imperfectly formed).

Electromagnetic Time Reversal: Application to Electromagnetic Compatibility and Power Systems,
First Edition. Edited by Farhad Rachidi, Marcos Rubinstein and Mario Paolone.
© 2017 John Wiley & Sons, Ltd. Published 2017 by John Wiley & Sons, Ltd.
Companion Website: www.wiley.com/go/rachidi55

of an obvious reality to everyone, there is nothing harder than giving a definition that truly captures its essence using concepts other than our intuitive idea of time itself. Let us look at some entries given in dictionaries:

> Indefinite, unlimited duration in which things are considered as happening in the past, present, or future; every moment there has ever been or ever will be.
>
> *(Webster's New World College Dictionary)*

> The indefinite continued progress of existence and events in the past, present, and future regarded as a whole.
>
> *(Oxford Dictionaries)*

Etienne Klein, a French physicist, has also proposed an interesting definition for time [1]: a jail on wheels. Why a jail? Because we are not free to choose our position along the timeline. We are in the present instant and we cannot get away from it. Why on wheels? Because time moves forward. *Time* takes us from the present to the future.

One can find hundreds of other definitions, proposed by philosophers, physicists, and linguists. Many of them use metaphors to describe this concept, as in the one proposed by Klein. However, none of these tells us about the nature of *time* since some idea of *time* is used in the definition itself [1]. This paradox was noticed by Saint Augustine in the fourth century: "If I am not asked, I know what time is; but if I am asked, I do not."

Physicists have managed to consider *time* as an operative concept. The first mathematical expression of physical time was enunciated by Galileo and formalized by Newton, assuming that time has one dimension and is expressed by a real number. Furthermore, in that definition, time is absolute in the sense that there is only one time associated with any given moment, and that time is the same everywhere in the universe. In 1905, Einstein's special relativity theory showed that time is not an absolute quantity and should be considered in relation to space. Quantum mechanics changed again the time paradigm [2] with the Heisenberg uncertainty principle, which applies not only to position and momentum, but also to time and energy.

1.3   Time Reversal or Going Back in Time

In this section, using a simple example, we will present three approaches that can be used to effectively make a system go back in time, in the sense that it retraces the path it came from in the immediate past. The three methods that will be discussed are (i) the recording of the state of the system throughout its evolution from time $t = 0$ to a time $t = t_1$, (ii) the use of expressions that describe the evolution of the system as a function of time, and (iii) the imposition of initial conditions so that the system regresses, following its own natural defining equations, towards the states it went through in its past. We will also discuss the conditions under which each one of the approaches can be applied and the implications for practical applications.

The simple example we will consider is that of an object of a given mass launched at an angle near the surface of the earth, as shown in Figure 1.1. The resistance of the air is considered to be zero. We will concentrate on the position of the object as the physical quantity of interest. Since the position can be represented pictorially, the success of an approach will be illustrated by the possibility of producing a movie of the trajectory of the object in reverse.

Note that, since we are using classical mechanics, we are using the Newtonian definition of time and not that of Einstein's relativity.

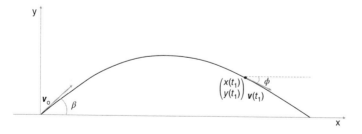

Figure 1.1 Projectile launched at an angle over a flat earth. The speed and position at time t_1 after the launch are shown after the highest point in the trajectory.

1.3.1 Approach 1: Recording of the State of the System Throughout Its Evolution

If the position of the object is recorded using, for instance, calibrated, synchronized video cameras, then the evolution of the system can be viewed both in the forward and, by reversing the order of the frames, in reverse. Clearly, the only condition that needs to be imposed for this approach to be applicable is that the physical quantity of interest be observable in principle at all times, although a limited number of samples may be sufficient depending on the intended application. No conditions need to be imposed on the properties of the underlying physical equations.

1.3.2 Approach 2: Use of Expressions that Describe the Evolution of the System as a Function of Time

The current state of knowledge in physics, based on observation, experimentation, and the application of the appropriate mathematical tools, allows us in a number of fields to write expressions that can be used to predict as a function of time, with known accuracy, the values of physical quantities associated with bounded physical systems.

Using classical mechanics, we can write the following function in Cartesian coordinates to describe the position of an object with a constant gravitational acceleration g and neglecting any friction with the air:

$$\begin{pmatrix} x(t) \\ y(t) \end{pmatrix} = \begin{pmatrix} x_0 + v_{x_0}(t - t_0) \\ y_0 + v_{y_0}(t - t_0) - \dfrac{1}{2}g(t - t_0)^2 \end{pmatrix} \tag{1.1a}$$

in which x_0 and y_0 represent the initial x and y coordinates of the object, v_{x_0} and v_{y_0} are the components of the initial velocity of the object and t_0 is the reference time for the launch of the object.

For the particular example of Figure 1.1, Equation (1.1a) can be written as

$$\begin{pmatrix} x(t) \\ y(t) \end{pmatrix} = \begin{pmatrix} |v_0| \cos(\beta)t \\ |v_0| \sin(\beta)t - \dfrac{1}{2}gt^2 \end{pmatrix} \tag{1.1b}$$

where x_0 and y_0 were set to zero since the projectile is launched from the origin of the coordinate system. Also, time is counted from $t = 0$ and, therefore, $t_0 = 0$. The speed can also be found as a function of time using Equation (1.2), which is simply the derivative of (1.1):

$$\begin{pmatrix} v_x(t) \\ v_y(t) \end{pmatrix} = \begin{pmatrix} |v_0| \cos(\beta) \\ |v_0| \sin(\beta) - gt \end{pmatrix} \tag{1.2}$$

If we let t increase from 0 to t_1, Equation (1.1) literally allows us to draw the trajectory of the projectile the same way we would observe it in a physical setup under the conditions posed above. Since the position is calculable at any time, it is possible to create a movie by drawing a point at the appropriate location in each of the movie frames based on the desired number of frames per second and on Equation (1.1). Of course, we could now stop the movie while the projectile is in flight and we could run it in reverse. We would then see the particle fly back to the point from where it was launched.

This approach can be used to produce movies that simulate forward-in-time and backward-in-time mechanical movement in more complex scenarios, as long as we are able to write the equations of movement for the complete time period of interest.

1.3.3   Approach 3: System Regressing Through Its Own Natural Defining Equations

In the first two approaches, the prior states of the system are found by looking at times in the past. In this approach, *the previous behavior of the system is reproduced in the future*. To achieve this, appropriate initial conditions are imposed on the original system so that it begins to retrace its previous states. An advantage of this approach is that there is no need to record the behavior of the system for the whole time interval of interest, since only the final conditions are required (they are the basis for the determination of the required new initial conditions for the system to regress). A disadvantage of this approach is the fact that the underlying physical equations of the particular system have to satisfy conditions that we will investigate later on, after we have illustrated the approach using the projectile example of

Figure 1.1. Another advantage of this approach, and the reason why we will concentrate on it in the remainder of the chapter, is that this is the only approach that can be applied both by simulation and experimentally.

Let us assume that we have measured the speed and the position of the projectile at time t_1 after the launch as illustrated in Figure 1.1.

Using (1.1) and (1.2), we can get the position and the speed that would be measured at time t_1:

$$\begin{pmatrix} x(t_1) \\ y(t_1) \end{pmatrix} = \begin{pmatrix} |v_0| \cos(\beta)t_1 \\ |v_0| \sin(\beta)t_1 - \frac{1}{2}gt_1^2 \end{pmatrix} \tag{1.3}$$

$$\begin{pmatrix} v_x(t_1) \\ v_y(t_1) \end{pmatrix} = \begin{pmatrix} |v_0| \cos(\beta) \\ |v_0| \sin(\beta) - gt_1 \end{pmatrix} \tag{1.4}$$

Using the approach described here, after having halted the evolution of the system at $t = t_1$, we now use the same system with appropriately modified initial conditions as follows.

The initial position for the regressing projectile will be the position of the object at time t_1 given by the right-hand side of (1.3):

$$\begin{pmatrix} x'_0 \\ y'_0 \end{pmatrix} = \begin{pmatrix} |v_0| \cos(\beta)t_1 \\ |v_0| \sin(\beta)t_1 - \frac{1}{2}gt_1^2 \end{pmatrix} \tag{1.5}$$

The new initial speed is given by the additive inverse[2] of the speed of the object at time t_1, which can be obtained by multiplying the right-hand side of (1.4) by -1:

$$\begin{pmatrix} v'_{x_0} \\ v'_{y_0} \end{pmatrix} = \begin{pmatrix} -|v_0| \cos(\beta) \\ -|v_0| \sin(\beta) + gt_1 \end{pmatrix} \tag{1.6}$$

To see if these initial conditions make the system behave in such a way that the future trajectory, starting at $t = t_1$, will actually retrace the path that the object had followed from $t = 0$ to

---

2 Note that we are reversing the direction of the speed vector based only on our intuition. A more rigorous procedure to determine the initial conditions will be discussed later on.

$t = t_1$, let us plug (1.5) and (1.6) into the underlying equations for the position of the object in the system, given by (1.1a):

$$
\begin{pmatrix} x(t) \\ y(t) \end{pmatrix}
$$

$$
= \begin{pmatrix} |v_0| \cos(\beta) t_1 - |v_0| \cos(\beta)(t - t_1) \\ |v_0| \sin(\beta) t_1 - \frac{1}{2} g t_1^2 + (-|v_0| \sin(\beta) + g t_1)(t - t_1) - \frac{1}{2} g(t - t_1)^2 \end{pmatrix}
$$

$$(1.7)$$

If Equation (1.7) indeed represents the reversion of the system starting at time t_1, then the position at time $2t_1$ should be the origin of the coordinate system, a fact that can be readily verified by introducing $t = 2t_1$ in (1.7).

Let us now investigate the conditions that need to be satisfied by the equations of the system so that it is possible to make it behave in a time-reversed manner after a selected point in time, which, as in the example above, we will call t_1.

Figure 1.2a represents a plot of the evolution of a physical quantity $f(t)$ as a function of time in a system. The quantity could be the position of a particle along a particular axis, the temperature, one of the components of the electric field intensity, or any other observable. In Figure 1.2a, the system has been frozen in time at $t = t_1$. We would like for the system's physical quantity to retrace the same values it just exhibited prior to t_1. The desired resulting behavior is shown by the dashed line in Figure 1.2b. The dashed curve can be obtained from the original function $f(t)$ by first translating it by t_1 to the left, then applying the T-symmetry transformation $T : t \rightarrow -t$, and finally translating it back to the right by t_1. The resulting function is $g(t) = f(-t + 2t_1)$.

From Figure 1.2, it can be concluded that the condition that needs to be satisfied by the underlying physical equations is that, given a solution $f(t)$ to the underlying differential equations governing a physical phenomenon, $g(t) = f(-t + 2t_1)$ must also be a solution. In the next section, we will formally use this condition to define time-reversal invariability in the strict sense.

1.3.4 Time Reversal in the Strict Sense and in the Soft Sense

We concluded the last section with a statement of the condition that the underlying equations governing the behavior of a

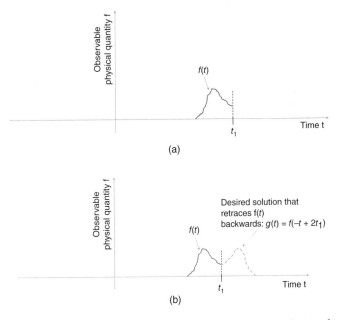

Figure 1.2 Plot of the evolution of the physical quantity f. (a) Evolution of $f(t)$ up to a time t_1. (b) Function $f(t)$ followed by the desired behavior $g(t)$, the dashed curve, of the system after t_1.

physical quantity must satisfy so that the system can be made to retrace the path followed in the immediate past by setting its initial conditions and letting it evolve into the future. We call this property of a physical system *time-reversal invariance in the strict sense*, as opposed to *time-reversal invariance in the soft sense*, which will be introduced later on in this section. The condition for time-reversal invariance in the strict sense is:

Strict Time-Reversal Invariance Condition

A system is time-reversal invariant with respect to a physical quantity in the strict sense if, given a solution $f(t)$ of its underlying equations, the time-reversed function $g(t)$, given by Equation (1.8) hereunder, is also a solution:

$$g(t) = f(-t + 2t_1) \tag{1.8}$$

The time t_1 is the time at which we freeze the system state before letting it regress. For simplicity, we will replace twice that constant time by k. The parameter k could also be set to zero since one can shift the origin of the time coordinate so that the observation period starts at a time $t = -2t_1$ and freeze the time at time $t = 0$.

In the case of the example of Figure 1.1, the differential equation that governs the movement of the object satisfies the condition for time-reversibility in the strict sense, as we will show in what follows.

The equations of the movement of the projectile are given by

$$\begin{pmatrix} \dfrac{\partial^2 x(t)}{\partial t^2} \\ \dfrac{\partial^2 y(t)}{\partial t^2} \end{pmatrix} = \begin{pmatrix} 0 \\ -g \end{pmatrix} \tag{1.9}$$

Assuming that $x(t)$ and $y(t)$ satisfy (1.9), we will now show that $x(-t + k)$ and $y(-t + k)$ also satisfy the equation:

$$\begin{pmatrix} \dfrac{\partial^2 x(-t + k)}{\partial t^2} \\ \dfrac{\partial^2 y(-t + k)}{\partial t^2} \end{pmatrix} = \begin{pmatrix} 0 \\ -g \end{pmatrix} \tag{1.10}$$

Note now that, since the right-hand side of (1.9) is a constant vector, its left-hand side must also be independent of time. Therefore, the left-hand sides of (1.9) and (1.10) must be equal since (1.10) is just (1.9) evaluated at time $-t + k$. The example in Figure 1.1 is thus time-reversal invariant in the strict sense.

When we originally wrote Equation (1.7), we used our intuition when, by inspection of the specific problem of a projectile in flight, the speed vector direction at $t = t_1$ was reversed and the appropriate initial conditions were selected and imposed. According to the developments presented here, it is possible to obtain Equation (1.7) by plugging Equation (1.1b) (as $f(t)$) into Equation (1.8). We will not include the details here but the result is indeed identical to (1.7).

The equations governing the example discussed up to now involved additive terms containing (1) the second derivative of the quantity of interest (which was the position) and (2) constant terms (see, for instance, Equation (1.9)). As we will see, the presence of other types of terms may violate the conditions for strict reversibility. We will now investigate time reversal in a system whose underlying equations involve such terms. This will lead to the definition of time-reversal invariability in the soft sense.

Let us consider now an object that is subjected to a force that is proportional to the elapsed time. For simplicity, we will assume the force's proportionality constant with respect to time to be m, the value of the mass of the object. Under these conditions, Newton's force equation for the object can be written as

$$a(t) = \frac{d^2x(t)}{dt^2} = t \tag{1.11}$$

We will now find a solution of (1.11) and test its time-reversibility.

The speed and the position are obtained by two successive integrations to be

$$v(t) = \frac{t^2}{2} + v_0 \tag{1.12}$$

$$x(t) = \frac{t^3}{6} + v_0 t + x_0 \tag{1.13}$$

Let us assume, to facilitate the calculations, that $v_0 = x_0 = 0$. With these assumptions, (1.12) and (1.13) can be written as

$$v(t) = \frac{t^2}{2} \tag{1.14}$$

$$x(t) = \frac{t^3}{6} \tag{1.15}$$

Let us now see if the function $g(t) = x(-t + 2t_1)$ satisfies Equation (1.11). First, let us write $g(t)$,

$$g(t) = x(-t + 2t_1) = \frac{(-t + 2t_1)^3}{6} \tag{1.16}$$

Plugging (1.16) into (1.11), and checking if the left- and right-hand sides are equal, we get

$$\frac{d^2\left[\frac{(-t + 2t_1)^3}{6}\right]}{dt^2} = -t + 2t_1 \neq t \tag{1.17}$$

The equality does not hold unless we apply the time-reversal transformation also to the right hand side of (1.11).

By applying the T-symmetry + translation transformation $t \rightarrow -t + 2t_1$ to the acceleration on the right-hand side of (1.11) we are modifying the behavior of the force in this example.

Indeed, the force, equal to mt in the original system in which the object evolved up to $t = t_1$, becomes $m(-t + 2t_1)$ after that time. It is therefore no longer sufficient to choose the appropriate initial conditions for the system to evolve into the future by retracing the original trajectory as if time were moving backwards. It is also necessary to change the behavior of the force after t_1 so that its functional dependence is $m(-t + 2t_1)$. In other words, the time-reversal transformation $T : t \rightarrow -t$ is also applied to the time-dependent terms that do not contain the physical quantity. This change is possible if the evolution of the system is carried out by simulation after $t = t_1$.

Phenomena that require changes in their defining equations such as the ones described here in order for their state to retrace in the future the path it followed in its past are referred to as *time-reversal invariant in the soft sense.*

Soft Time-Reversal Invariance Condition

A system is time-reversal invariant with respect to a physical quantity in the soft sense if, given a solution $f(t)$ of its underlying equations, the time-reversed function $g(t)$, given by Equation (1.18) hereunder, is also a solution of the underlying equation after appropriate changes to the equation are applied:

$$g(t) = f(-t + 2t_1) \tag{1.18}$$

The basic equations of electricity and magnetism, as well as equations in quantum mechanics appear to be time-reversal invariant in the soft sense, as will be shown in the next subsections.

1.3.5 Schrödinger Equation

Let us consider the Schrödinger equation in quantum mechanics that describes the wave nature of particles:

$$j\hbar \frac{\partial \psi(r, t)}{\partial t} = -\frac{\hbar^2}{2m} \nabla^2 \psi(r, t) + V(r, t)\psi(r, t) \tag{1.19}$$

where j is the square root of -1, $\hbar$ is the reduced Planck constant, obtained by dividing h by 2π, m is the particle's reduced mass, $V(r, t)$ its potential energy and $\psi(r, t)$ the wave function.

It can be readily seen that replacing t by $t' = -t + 2t_1$, the term on the left-hand side changes sign and, therefore, $\psi(r, -t + 2t_1)$ is not a solution of the Schrödinger equation. However, time-reversal symmetry is achieved when substituting t by $t' = -t + 2t_1$ and taking the complex conjugate of Equation (1.19):

$$j\hbar\frac{\partial \psi^*(r, t')}{\partial t'} = -\frac{\hbar^2}{2m}\nabla^2\psi^*(r, t') + V(r, t')\psi^*(r, t') \qquad (1.20)$$

where the term on the left-hand side remains unchanged since two sign changes, one due to the change of variables variables $t' = -t + 2t_1$ and another to the conjugate transformation $j^* = -j$, cancel.

Equation (1.20) shows that if $\psi(r, t)$ is a solution of the Schrödinger equation, then $\psi^*(r, -t + 2t_1)$ is also a solution. Therefore, the Schrödinger equation is invariant under time reversal in the soft sense because of the applied conjugate complex operation.[3]

1.3.6 Maxwell's Equations

Let us consider classical electromagnetism, in which the electric and magnetic fields obey Maxwell's equations:

$$\nabla \times \mathbf{E}(\mathbf{r}, t) = -\mu(\mathbf{r})\frac{\partial \mathbf{H}(\mathbf{r}, t)}{\partial t} \qquad (1.21)$$

$$\nabla \times \mathbf{H}(\mathbf{r}, t) = \varepsilon(\mathbf{r})\frac{\partial \mathbf{E}(\mathbf{r}, t)}{\partial t} + \mathbf{J}(\mathbf{r}, t) \qquad (1.22)$$

$$\nabla \cdot (\varepsilon(\mathbf{r})\mathbf{E}(\mathbf{r}, t)) = \rho(\mathbf{r}, t) \qquad (1.23)$$

$$\nabla \cdot (\mu(\mathbf{r})\mathbf{H}(\mathbf{r}, t)) = 0 \qquad (1.24)$$

Let us express the current density as the sum of two parts, one corresponding to a source, imposed by an external agent, and one being the consequence of a non-zero conductivity in the medium:

$$\nabla \times \mathbf{E}(\mathbf{r}, t) = -\mu(\mathbf{r})\frac{\partial \mathbf{H}(\mathbf{r}, t)}{\partial t} \qquad (1.25)$$

---

3 It should be noted that only $|\psi^2(r, t)|$ leads to observable effects [3].

$$\nabla \times \mathbf{H}(\mathbf{r}, t) = \varepsilon(\mathbf{r})\frac{\partial \mathbf{E}(\mathbf{r}, t)}{\partial t} + \mathbf{J}_s(\mathbf{r}, t) + \sigma(\mathbf{r})\mathbf{E}(\mathbf{r}, t) \qquad (1.26)$$

$$\nabla \cdot (\varepsilon(\mathbf{r})\mathbf{E}(\mathbf{r}, t)) = \rho(\mathbf{r}, t) \qquad (1.27)$$

$$\nabla \cdot (\mu(\mathbf{r})\mathbf{H}(\mathbf{r}, t)) = 0 \qquad (1.28)$$

Let us now assume that we have found solutions $\mathbf{E}(\mathbf{r}, t)$ and $\mathbf{H}(\mathbf{r}, t)$ for the electric and the magnetic fields, respectively. We are interested in the conditions under which the time-reversed functions $\mathbf{E}(\mathbf{r}, -t + 2t_1)$ and $\mathbf{H}(\mathbf{r}, -t + 2t_1)$ satisfy Maxwell's equations.

Let us write the same equations for the time-reversed functions. Note that we have added a question mark at the end of each equation to indicate that we are, at this time, not sure that the equations are satisfied.

$$\nabla \times \mathbf{E}(\mathbf{r}, -t + 2t_1) = -\mu(\mathbf{r})\frac{\partial \mathbf{H}(\mathbf{r}, -t + 2t_1)}{\partial t}? \qquad (1.29)$$

$$\nabla \times \mathbf{H}(\mathbf{r}, -t + 2t_1) = \varepsilon(\mathbf{r})\frac{\partial \mathbf{E}(\mathbf{r}, -t + 2t_1)}{\partial t} + \mathbf{J}_s(\mathbf{r}, -t + 2t_1)$$
$$+ \sigma(\mathbf{r})\mathbf{E}(\mathbf{r}, -t + 2t_1)? \qquad (1.30)$$

$$\nabla \cdot (\varepsilon(\mathbf{r})\mathbf{E}(\mathbf{r}, -t + 2t_1)) = \rho(\mathbf{r}, -t + 2t_1)? \qquad (1.31)$$
$$\nabla \cdot (\mu(\mathbf{r})\mathbf{H}(\mathbf{r}, -t + 2t_1)) = 0? \qquad (1.32)$$

Note also that, in (1.30) and (1.31), we also applied the T-symmetry transformation to the sources (as discussed in Section 1.3.4).

Let us pose $t' = -t + 2t_1$ and rewrite Equations (1.29) to (1.32) as a function of t':

$$\nabla \times \mathbf{E}(\mathbf{r}, t') = \mu(\mathbf{r})\frac{\partial \mathbf{H}(\mathbf{r}, t')}{\partial t'} \qquad (1.33)$$

$$\nabla \times \mathbf{H}(\mathbf{r}, t') = -\varepsilon(\mathbf{r})\frac{\partial \mathbf{E}(\mathbf{r}, t')}{\partial t'} + \mathbf{J}_s(\mathbf{r}, t') + \sigma(\mathbf{r})\mathbf{E}(\mathbf{r}, t') \qquad (1.34)$$

$$\nabla \cdot (\varepsilon(\mathbf{r})\mathbf{E}(\mathbf{r}, t')) = \rho(\mathbf{r}, t') \qquad (1.35)$$

$$\nabla \cdot (\mu(\mathbf{r})\mathbf{H}(\mathbf{r}, t')) = 0 \qquad (1.36)$$

Since the first derivatives will change the sign of the terms involved, we can see that Equations (1.33) and (1.34), as they are shown here, are not time-reversal invariant. Thus, the transformation $t \rightarrow -t + 2t_1$ does not satisfy time reversibility alone. However, by inspection, we can see that changing the sign of the magnetic field, of the source current density, and of the conductivity will result in the time reversibility of the Maxwell equations.[4]

A discussion is in order on the necessity of the sign change for the magnetic field and the current density, which has created some controversy in the literature (see, for example, [4] to [7]). Indeed, it was argued by Albert [6] that Maxwell's equations are not time-reversal invariant and the sign change of the magnetic field was just a mathematical trick to save the time-reversal invariance (see the discussion in [5]). Snieder [3] argued that there is a physical reason for which the electric current density should change sign under time reversal. When the direction of time is reversed, the velocity of the charges changes sign. As a result, the associated current density should change sign as well. Finally, a change of sign of the current density should result in a change of sign of the associated magnetic field.

The fact that changes are required in the equations means that Maxwell's equations are time-reversal invariant in the soft sense.

1.3.7 Time-Reversal Process: Summary

We have seen that in order to be able to reproduce the past behavior of a system in the future, the condition that needs to be satisfied by the underlying physical equations is that, given a solution $f(t)$ to the underlying differential equations related to a physical phenomenon, $g(t) = f(-t + k)$ must also be a solution. Since k is a constant related to the arbitrary reference origin for the time, this condition can be simply checked by verifying that the underlying equations remain unchanged by making the T-symmetry substitution $t \rightarrow -t$. If this condition is satisfied, the physical law described by the equations is called invariant under time reversal in the strict sense. We have also seen that this

---

4 A second possibility is the change the signs of the electric field, the conductivity and the charge density.

condition alone might not be enough to satisfy time reversibility. For the case of Maxwell's equations, for instance, the sign of the magnetic field, of the source current density, and of the conductivity should also be changed. We have defined phenomena that require these additional conditions as time-reversal invariant in the soft sense.

It is interesting to observe that most laws of nature feature invariance under time reversal, either in the strict or in the soft sense. There are a few exceptions, an example of which is the weak force governing radioactive decay [3].

1.4 Application of Time Reversal in Practice

Assume a steel ball is propelled at a time $t = 0$ into the playfield of a pinball machine [8]. The ball will follow a complex trajectory moving in different directions, striking different targets and bumping at kickers and slingshots. Now, let us apply the third approach presented in Section 1.3.3 and let us freeze the event at a given time t and apply the time-reversal process as described in the previous section. The ball will then converge back towards the plunger, from which it was originally launched, as if we were watching the film of the game backward. Interestingly, this thought experiment is in reality infeasible in practice, even using computer simulations due to the general impossibility of representing real numbers with a finite number of bits.[5] After a few collisions, the ball will miss a barrier it should have hit and the rest of its route will be changed. As a consequence, the ball will never reach back to its origin. The reason why time reversal is practically impossible to realize in dynamical systems is the deterministic chaos, namely the fact that these systems are very sensitive to the initial conditions. Small differences in the initial conditions, in this case the measurement of the position and speed at time t, would result in a diverging behavior of the system. The perturbations in the solution in chaotic dynamical systems grow exponentially with time, following a function $\exp(\mu t)$, where μ is the so-called Lyapounov exponent [3].

---

5 Note that it would be possible to make the ball retrace the path to the launch position if the complete process of forward and back-propagation were made using idealized analytical expressions for Newtonian mechanics.

Furthermore, the presence of friction breaks the time-reversal symmetry of the laws of motion.

Unlike dynamical systems, the application of the time-reversal process to wave phenomena appears to be very effective. Fields of waves are not very sensitive to the initial conditions. Perturbations in the solution in wave systems grow at a much smaller rate compared to dynamical systems [3]. Intuitively, this can be explained by the fact that waves travel along all possible scattering paths while particles travel along a unique path [3]. Furthermore, the dispersion effect is, in many practical situations, negligible in wave phenomena. As a result, and following the pioneering work of Fink and co-workers (e.g., [9]), the time-reversal process has found a great number of applications, first in acoustics and later in electromagnetics.

In the next section, we will discuss the use of the time-reversal process as a way of refocusing electromagnetic fields towards their source. Note that in the remainder of this chapter, the time reversal process is defined according to the third approach presented in Section 1.3.3.

1.5 Refocusing of Electromagnetic Waves Using Time Reversal

1.5.1 Time-Reversal Cavity

The concept of a time-reversal cavity was proposed by Cassereau and Fink [10] for acoustic waves and later extended to electromagnetic waves, using the Lorentz reciprocity principle [11]. A time-reversal cavity extends the concept of a time-reversal mirror (TRM) [9, 12], a technique used to refocus an emitted wave back to its source.

Consider the situation depicted in Figure 1.3a, in which a source emits an impulsive electromagnetic field in a linear and non-magnetic medium.[6] Consider a closed surface S surrounding the source and suppose that we are able to determine the tangential fields generated by the source at any point on this surface.

------

6 The medium can be inhomogeneous.

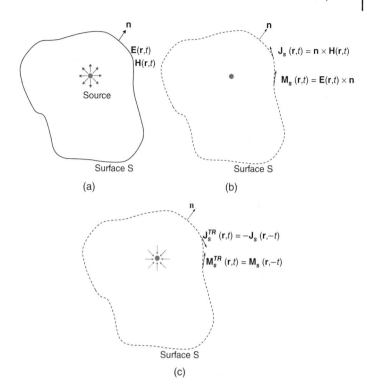

Figure 1.3 (a) A source emits an impulsive electromagnetic field in the medium at time $t = 0$. We determine the tangential fields generated by the source at any point on the closed surface S as a function of time.
(b) Making use of the equivalence theorem, it is possible to replace the source in (a) by equivalent electric and magnetic current sources $\mathbf{J}_s$ and $\mathbf{M}_s$.
(c) Time-reversing the equivalent sources on the surface will result in time-reversing the electromagnetic fields in all the medium within that surface, converging back to the source location.

Making use of the equivalence theorem, it is possible to replace the source in Figure 1.3a by equivalent electric and magnetic current sources $\mathbf{J}_s$ and $\mathbf{M}_s$ on the surface S (Figure 1.3b), which are given by

$$\mathbf{J}_s(\mathbf{r}, t) = \mathbf{n} \times \mathbf{H}(\mathbf{r}, t) \tag{1.37}$$

$$\mathbf{M}_s(\mathbf{r}, t) = \mathbf{E}(\mathbf{r}, t) \times \mathbf{n} \tag{1.38}$$

The fields outside the surface can be calculated either by computing the fields from the original source in Figure 1.3a or by calculating the fields from the new electric and magnetic currents on S, shown in Figure 1.3b. In other words, according to the equivalence theorem, the fields generated by the equivalent sources on the surface will be the same as the fields generated by the original source.

In addition, according to the uniqueness theorem, the specification of the electric and/or magnetic field on the surface S corresponds to a unique distribution of the electromagnetic field in the volume surrounded by S.

Now assume that we time-reverse the electric and magnetic fields in all the points of the volume surrounded by S, namely

$$\mathbf{E}(\mathbf{r}, t) \rightarrow \mathbf{E}(\mathbf{r}, -t) \tag{1.39}$$

$$\mathbf{H}(\mathbf{r}, t) \rightarrow -\mathbf{H}(\mathbf{r}, -t) \tag{1.40}$$

In this case, the new equivalent sources on the surface S become

$$\mathbf{J}_s^{TR}(\mathbf{r}, t) = -\mathbf{n} \times \mathbf{H}(\mathbf{r}, -t) \tag{1.41}$$

$$\mathbf{M}_s^{TR}(\mathbf{r}, t) = \mathbf{E}(\mathbf{r}, -t) \times \mathbf{n} \tag{1.42}$$

Obviously, the new equivalent sources corresponding to the time-reversed electric and magnetic fields are time-reversed versions of the direct-time sources (1.37) and (1.38):

$$\mathbf{J}_s^{TR}(\mathbf{r}, t) = -\mathbf{J}_s(\mathbf{r}, -t) \tag{1.43}$$

$$\mathbf{M}_s^{TR}(\mathbf{r}, t) = \mathbf{M}_s(\mathbf{r}, -t) \tag{1.44}$$

Likewise, time-reversing the equivalent sources on the surface will result in time-reversing the electromagnetic fields in all the medium within that surface (Figure 1.3c).

Starting from this observation, the concept of a time-reversal cavity allows refocusing a wave back to its source, using a two-step process. In the first step, the electromagnetic fields generated by the source are determined over the surface S forming the cavity. In the second step, the internal source is removed from the medium and equivalent time-reversed sources on the surface are considered. The resulting field in the second step will be a time-reversed copy of the fields in the first step and thus will converge back to the source, before diverging again. In order to eliminate the diverging field, the initial source must be replaced by a time-reversed sink [12].

1.5.2 Use of a Limited Number of Sensors

Obviously, a time-reversal cavity cannot be realized experimentally [10] because it requires an infinite number of transducers covering a closed surface around the medium to obtain information about all wavefronts propagating in all directions [13]. In practice, the fields can be measured using a limited number of sensors and the question arises whether the focusing property of time reversal remains intact. Draeger *et al.* [13, 14] have shown that, in the case of acoustic waves, focusing is possible using a single element in a closed reflecting cavity with negligible absorption. The same focusing property was obtained with electromagnetic waves (e.g., [15]). Derode *et al.* [16] have shown that, compared to a homogeneous medium, a higher focusing quality can be achieved in an inhomogeneous medium, as a result of multiple reflections and scattering.

Consider the example of a source that emits an electromagnetic field impulse in free space (triangle at the center of the cylindrical wavefront in Figure 1.4a). This could be, for example, a cloud-to-ground lightning discharge. Suppose we have three sensors that record the fields generated by the source (shown in Figure 1.4a on the top, left and right parts of the domain). Figure 1.4a shows the cylindrical wave generated by the source at a given time. Time-reversing the received waveforms captured by each sensor and re-emitting them back into the medium will result in a maximum peak field at the location of the source, at which the injected time-reversed waveforms contributing to the total field are in phase (Figure 1.4b).

Now let us consider the case where the the medium contains two perfectly conducting reflecting walls, as shown in Figure 1.5a [17]. In this case, it will be possible to locate the source with only one sensor. Indeed, the source point (triangle at the center of the cylindrical wavefront) generates a field, which is reflected by the two walls (Figure 1.5a), which, by virtue of image theory, can be replaced by three mirrored sources. Hence the sensor (triangle shown on the left part of the domain) receives four successive

$t = 2.3$

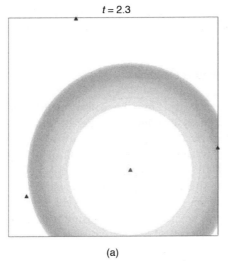

(a)

$t = 0$

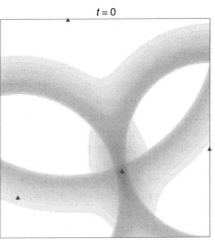

(b)

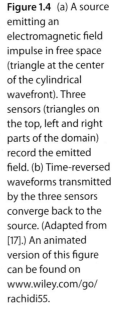

Figure 1.4 (a) A source emitting an electromagnetic field impulse in free space (triangle at the center of the cylindrical wavefront). Three sensors (triangles on the top, left and right parts of the domain) record the emitted field. (b) Time-reversed waveforms transmitted by the three sensors converge back to the source. (Adapted from [17].) An animated version of this figure can be found on www.wiley.com/go/rachidi55.

waves. In the second step, the sensor re-emits the time-reversed fields into the medium, which are in turn reflected by the two walls (Figure 1.5b). As can be seen in Figure 1.5b, the maximum amplitude field is obtained at the source location and, therefore, a single sensor makes it possible to locate the source in this case.

Figure 1.5 (a) A source (triangle at the center of the cylindrical wavefront) emitting an electromagnetic field impulse in a medium containing two reflecting walls (bottom and right). One sensor (triangle on the left part of the domain) records the emitted field. (b) Time-reversed waveform received by the sensor converges back to the source. (Adapted from [17].) An animated version of this figure can be found on www.wiley.com/go/rachidi55.

$t = 2.6$

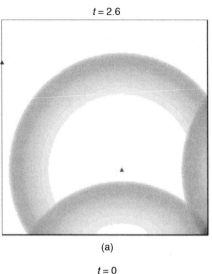

(a)

$t = 0$

(b)

Another important point to be mentioned is that the field distribution resulting from the injection of the time-reversed fields from a finite number of sources back into the medium will obviously not be a time-reversed copy of the fields in direct time [18, 19].

1.5.3 Time Reversal and Matched Filtering

Time reversal is similar to the concept of matched filtering used in signal processing. Consider, as an example, an antenna A in a linear, time-invariant medium. A signal $x(t)$ is applied to antenna A. Suppose that we have N receiving antennas $B_1, ..., B_N$. The signal received at the output of antenna B_i can be written as

$$y_i(t) = x(t) \otimes h_i(t) \tag{1.45}$$

where $h_i(t)$ is the system impulse response and $\otimes$ denotes the convolution product. Now, let us time-reverse the received signals and feed each one of them back to its corresponding antenna B_i. Using reciprocity, we can express the signal received at antenna A as

$$z(t) = \sum_{i=1}^{N} y(-t) \otimes h_i(t) = \sum_{i=1}^{N} x(-t) \otimes h_i(-t) \otimes h_i(t)$$

$$= \sum_{i=1}^{N} R_{ii}(t) \otimes x(-t) \tag{1.46}$$

in which $R_{ii}(t)$ is the autocorrelation function of the system impulse response $h_i(t)$, the Fourier transform of which is given by

$$R_{ii}(\omega) = H_i(\omega) H_i^*(\omega) = |H_i(\omega)|^2 \tag{1.47}$$

Thus, the signal $z(t)$ received at antenna A has the same phase as the input signal $x(t)$, but is conjugated and its amplitude is modified by the autocorrelation function. Therefore, it cannot be considered as a perfect replica of the emitted signal.

1.6 Applications of Time Reversal in Electrical Engineering

The first experiment using time reversal in electromagnetism was reported by Bogert [20] in 1957, in which a time-reversal technique was used to compensate delay distortion on a slow-speed picture transmission system.

The time-reversal technique was popularized in the scientific community by Fink and his colleagues in the 1990s in various studies related to acoustics (e.g., [8] to [10], [13], [16], and [21]), and later to electromagnetism (e.g., [12] and [15]).

In the twenty-first century, the time-reversal technique has emerged as a very interesting technique with potential applications in various fields of engineering, leading to mature technologies with unprecedented performance compared to classical techniques. Applications of time reversal include

- Focusing and amplification of electromagnetic waves (e.g., [15] and [22])
- Biomedical engineering (e.g., [23])
- Imaging (e.g., [24] to [27])
- Wireless communications (e.g., [28])
- EMC testing (e.g., [29])
- Fault detection and location (e.g., [30] to [32])
- Earthquake detection [33]
- Landmine detection (e.g., [34] and [35])
- Communications and radar (e.g., [36])
- Lightning location [37, 38].

It is expected that the fields of application of electromagnetic time reversal (EMTR) will continue to grow in the near future. In the following chapters of this book, some of the applications of the electromagnetic time reversal to power systems and electromagnetic compatibility will be described.

References

1 E. Klein, *Chronos: How Time Shapes Our Universe*. New York: Avalon, 2005.

2 J. G. Cramer, "The plane of the present and the new transactional paradigm of time," in *Time and the Instant*, R. Drurie, ed. United Kingdom: Clinamen Press, 2001.

3 R. Snieder, "Time-reversal invariance and the relation between wave chaos and classical chaos," in *Imaging of Complex Media with Acoustic and Seismic Waves*, M. Fink, W. Kuperman, J.-P. Montagner, and A. Tourin, eds, vol. 84, pp. 1–16. Berlin and Heidelberg: Springer, 2002.

4 J. Earman, "What time reversal invariance is and why it matters," *International Studies in the Philosphy of Science*, vol. 16, pp. 245–264, 2002.

5 D. B. Malament, "On the time reversal invariance of classical electromagnetic theory," *Studies in History and Philosophy of Science Part B: Studies in History and Philosophy of Modern Physics*, vol. 35, pp. 295–315, 2004.

6 D. Z. Albert, *Time and Chance*. Cambridge, Mass.: Harvard University Press, 2009.

7 F. Arntzenius and H. Greaves, "Time reversal in classical electromagnetism," *The British Journal for the Philosophy of Science*, vol. 60, pp. 557–584, 2009.

8 M. Fink, "Time-reversed acoustics," *Scientific American*, vol. 281, no. 5, pp. 91–97, 1999.

9 M. Fink, "Time reversal of ultrasonic fields. I. Basic principles," *IEEE Transactions on Ultrasonics, Ferroelectrics and Frequency Control*, vol. 39, pp. 555–566, 1992.

10 D. Cassereau and M. Fink, "Time-reversal of ultrasonic fields. III. Theory of the closed time-reversal cavity," *IEEE Transactions on Ultrasonics, Ferroelectrics and Frequency Control*, vol. 39, pp. 579–592, 1992.

11 R. Carminati, R. Pierrat, J. d. Rosny, and M. Fink, "Theory of the time reversal cavity for electromagnetic fields," *Optics Letters*, vol. 32, pp. 3107–3109, 2007.

12 J. D. Rosny, G. Lerosey, and M. Fink, "Theory of electromagnetic time-reversal mirrors," *IEEE Transactions on Antennas and Propagation*, vol. 58, pp. 3139–3149, 2010.

13 C. Draeger and M. Fink, "One-channel time reversal of elastic waves in a chaotic 2D-silicon cavity," *Physical Review Letters*, vol. 79, pp. 407–410, 1997.

14 C. Draeger, J.-C. Aime, and M. Fink, "One-channel time-reversal in chaotic cavities: experimental results," *The Journal of the Acoustical Society of America*, vol. 105, pp. 618–625, 1999.

15 G. Lerosey, J. de Rosny, A. Tourin, A. Derode, G. Montaldo, and M. Fink, "Time reversal of electromagnetic waves," *Physical Review Letters*, vol. 92, p. 193904, 2004.

16 A. Derode, A. Tourin, and M. Fink, "Limits of time-reversal focusing through multiple scattering: long-range correlation,"

The Journal of the Acoustical Society of America, vol. 107, pp. 2987–2998, 2000.

17 G. Lugrin, "Locating transient disturbance sources and modeling their interaction with transmission lines. Use of Electromagnetic time reversal and asymptotic theory," PhD thesis, Swiss Federal Institute of Technology (EPFL), Lausanne, Switzerland, 2016.

18 C. Altman and K. Suchy, *Reciprocity, Spatial Mapping and Time Reversal in Electromagnetics*. Springer, 2011.

19 W. M. G. Dyab, T. K. Sarkar, A. Garc, L. a Rez, M. Salazar-Palma, *et al.*, "A critical look at the principles of electromagnetic time reversal and its consequences," *IEEE Antennas and Propagation Magazine*, vol. 55, pp. 28–62, 2013.

20 B. Bogert, "Demonstration of delay distortion correction by time-reversal techniques," *IRE Transactions on Communication Systems*, vol. 5, pp. 2–7, 1957.

21 F. Wu, J. L. Thomas, and M. Fink, "Time reversal of ultrasonic fields. Il. Experimental results," *IEEE Transactions on Ultrasonics, Ferroelectrics and Frequency Control*, vol. 39, pp. 567–578, 1992.

22 M. Davy, J. d. Rosny, and M. Fink, "Focusing and amplification of electromagnetic waves by time-reversal in an leaky reverberation chamber," in *Antennas and Propagation Society International Symposium, 2009. APSURSI '09. IEEE*, pp. 1–4, 2009.

23 M. Fink, G. Montaldo, and M. Tanter, "Time-reversal acoustics in biomedical engineering," *Annual Review of Biomedical Engineering*, vol. 5, pp. 465–497, 2003.

24 H. W. Chun, T. R. James, and C. Fu-Kuo, "A synthetic time-reversal imaging method for structural health monitoring," *Smart Materials and Structures*, vol. 13, p. 415, 2004.

25 G. Montaldo, D. Palacio, M. Tanter, and M. Fink, "The time reversal kaleidoscope: a new concept of smart transducers for 3D imaging," in *2003 IEEE Symposium on Ultrasonics*, vol. 1, pp. 42–45, 2003.

26 D. Liu, G. Kang, L. Li, Y. Chen, S. Vasudevan, W. Joines, *et al.*, "Electromagnetic time-reversal imaging of a target in a cluttered environment," *IEEE Transactions on Antennas and Propagation*, vol. 53, pp. 3058–3066, 2005.

27 W. Zhang, A. Hoorfar, and L. Li, "Through-the-wall target localization with time reversal MUSIC method," *Progress in Electromagnetics Research*, vol. 106, pp. 75–89, 2010.

28 P. Kyritsi, G. Papanicolaou, P. Eggers, and A. Oprea, "Time reversal techniques for wireless communications," in *Vehicular Technology Conference, 2004. VTC2004-Fall. 2004 IEEE 60th*, vol. 1, pp. 47–51, 2004.

29 A. Cozza and e.-A. Abd el-Bassir Abou, "Accurate radiation-pattern measurements in a time-reversal electromagnetic chamber," *IEEE Antennas and Propagation Magazine*, vol. 52, pp. 186–193, 2010.

30 L. Abboud, A. Cozza, and L. Pichon, "Utilization of matched pulses to improve fault detection in wire networks," in *9th International Conference on Intelligent Transport Systems Telecommunications (ITST), 2009*, pp. 543–548, 2009.

31 L. El Sahmarany, L. Berry, N. Ravot, F. Auzanneau, and P. Bonnet, "Time reversal for soft faults diagnosis in wire networks," *Progress in Electromagnetics Research M*, vol. 31, pp. 45–58, 2013.

32 R. Razzaghi, G. Lugrin, H. M. Manesh, C. Romero, M. Paolone, and F. Rachidi, "An efficient method based on the electromagnetic time reversal to locate faults in power networks," *IEEE Transactions on Power Delivery*, vol. 28, pp. 1663–1673, 2013.

33 C. Larmat, J.-P. Montagner, M. Fink, Y. Capdeville, A. Tourin, and E. Clévédé, "Time-reversal imaging of seismic sources and application to the great Sumatra earthquake," *Geophysical Research Letters*, vol. 33, 2006.

34 M. Alam, J. H. McClellan, P. D. Norville, and W. R. Scott Jr, "Time-reverse imaging for detection of landmines," in *Proceedings of SPIE 5415, Detection and Remediation Technologies for Mines and Minelike Targets IX*, vol. 167, September 21, 2004, doi: 10.1117/12.542686.

35 A. Sutin, P. Johnson, J. TenCate, and A. Sarvazyan, "Time reversal acousto-seismic method for land mine detection," in *Proceedings of SPIE 5794, Detection and Remediation Technologies for Mines and Minelike Targets X*, June 10, 2005, doi: 10.1117/12.609838.

36 H. Zhai, S. Sha, V. K. Shenoy, S. Jung, M. Lu, K. Min, *et al.*, "An electronic circuit system for time-reversal of ultra-wideband short impulses based on frequency-domain approach," *IEEE Transactions on Microwave Theory and Techniques*, vol. 58, pp. 74–86, 2010.

37 N. Mora, F. Rachidi, and M. Rubinstein, "Application of the time reversal of electromagnetic fields to locate lightning discharges," *Atmospheric Research*, vol. 117, pp. 78–85, 2012.

38 G. Lugrin, N. Mora Parra, F. Rachidi, M. Rubinstein, and G. Diendorfer, "On the location of lightning discharges using time reversal of electromagnetic fields," *IEEE Transactions on Electromagnetic Compatibility*, vol. 56, pp. 149–158, 2014.

2

Time Reversal in Diffusive Media

A. Cozza and F. Monsef

Group of Electrical Engineering Paris (GeePs), Gif-sur-Yvette, France

2.1 Introduction

Fog, clouds, matt surfaces, even the human skin. All of them present a strong diffusive behavior. Without it, the sky would be a dull black. When light propagating along a single direction interacts with these diffusive media, it is scattered over a large fan of directions, making it fundamentally impossible to understand where the light came from in the first place. The next time you drink a glass of milk observe it closely and ask yourself if you can guess where the light that brightens it comes from.

Diffusive media are very common at visible optical frequencies; our brains are so accustomed to them that it fundamentally exploits their signature response as one of the features it uses for classifying surfaces. Switching to acoustics, at audible frequencies we are more comfortable with diffusion than with reflections: just think about how annoying are the separate echoes in a train station and how much we enjoy a concert hall with "good acoustics," a more common name for perfect diffusion [1, 2].

Enter now microwave frequencies and the concepts introduced by decades of undergraduate courses. Free-space propagation sets the standard, with the odd diffraction coming

Electromagnetic Time Reversal: Application to Electromagnetic Compatibility and Power Systems, First Edition. Edited by Farhad Rachidi, Marcos Rubinstein and Mario Paolone.

from a few isolated scatterers presented more as a nuisance than an opportunity, because of their complexity. Single mirror reflections are typically presented as the building blocks for multipath propagation; wave diffusion is hardly mentioned.

However, diffusion is not only important in understanding propagation at optical frequencies. Diffusion also actually happens quite often with microwaves. As a matter of fact, non-line-of-sight propagation in rooms and urban canyons share some of the characteristics of wave diffusion, which is better known as the class of Rayleigh channels in telecommunication theory [3].

Perhaps more important is the fact that wave diffusion can be useful. Reverberation chambers are the champions of wave diffusion, as their peculiar characteristics heavily depend on the ability to turn coherent excitations into a chaos of waves propagating along (ideally) every possible direction at the same time. When used as the basis for time-reversal applications, diffusive media can be shown to take a whole new dimension that makes them suddenly appear as very appealing media for wave propagation, rather than the messy and uncontrollable medium we are used to expect with harmonic excitations.

This chapter presents a summary of the work done in applying time reversal to diffusive media. It is organized into three parts. First, the main features of diffusive media are presented, starting from their physical origin leading to the derivation of black-box statistical models; these properties will be fundamental in the derivation of the results presented later. Time-reversed excitations are then considered, first for the generation of fields and the transmission of signals. The powerful self-averaging enabled by diffusive media is demystified, proving why it can only happen with such media. Self-averaging powers all of the properties of time reversal in diffusive media, as its ability to generate well-polarized fields in a medium is considered incapable of doing it.

These results are then pushed a step further, by studying the generation of coherent wavefronts. Diffusive media are shown to be a potentially more effective solution for this task than free-space-based wave generators: a single antenna is shown to be capable of generating a large number of focusing wavefronts just by playing on the signals applied to it. Applications presented throughout the chapter are based on the use of reverberation chambers.

2.2 Fundamental Properties of Diffusive Media

Before discussing the applications of time reversal in diffusive media, it is necessary to have at least an intuitive understanding of the physical origin of their properties. These are described in Section 2.2.1, introducing the plane-wave spectrum representation. From these observations, general statistical properties are inferred and used in order to introduce in Section 2.2.2 statistical black-box representations that capture the macroscopic behavior of diffusive media.

2.2.1 Understanding and Describing Wave Propagation in Diffusive Media

The notion of a diffusive medium is associated with environments where waves are submitted to a large (ideally infinite) number of scattering events. Such media go fundamentally under three categories: (1) rough surfaces, where an impinging wave is randomly back-scattered; (2) complex distributions of small scatterers, through which an impinging wave travels, e.g., Rayleigh scattering; (3) multiple scattering of waves generated in a (at least) partially closed medium, e.g., indoor environments.

In the context of this chapter, only the third configuration will be considered. In fact, all of the three configurations could be described as discussed in this section and the next one. However, the applications presented later refer, at a certain point, to closed media where scattering events are weakly dissipative. This condition is necessary in order to ensure that waves scattered by a portion of the medium boundary will travel to another part of the boundaries and undergo another round of scattering events, and so on for a large number of times. A closed structure without the assumption of weak losses could not support strong diffusion, making the simplified models in Section 2.2.2 too crude.

The coherence of waves propagating in these media is thus broken into small regions of coherence: what would have been an extended wavefront in free space is reduced to a collection of cells moving randomly in all directions [4,5], an outcome known as speckle distribution. An example is shown in Figure 2.1.

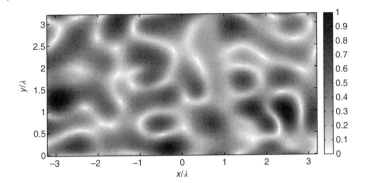

Figure 2.1 An example of spatial speckle (amplitude) distribution generated by a superposition of randomly excited plane waves. Nodal surfaces, shown in white, delimit coherent cells about a half-wavelength wide.

Multiple scattering does not necessarily imply that portions of the original energy propagate along every possible direction; media such as tunnels with cross-sections covering several wavelength present diffusion, but mainly across a reduced set of directions, since the average energy it conveys moves longitudinally.

The simplest representation that can be picked up for all of these configurations is the plane-wave spectrum [6]. It is widely used across all disciplines involving wave propagation, from acoustics to optics and even in quantum mechanics, taking a variety of names, as reciprocal-space representation or Fourier optics. In fact, it is just an extension of Fourier transform, applied to space rather than time.

The following definition of the plane-wave spectrum $\tilde{E}(k, v)$ will be used:

$$E(r, v) = \int dk \, \tilde{E}(k, v) e^{-j k \cdot r} \tag{2.1}$$

to represent a field distribution $E(r, v)$, where k is the propagation vector of each plane wave on which the spatial field distribution is expanded and v is the frequency at which the plane-wave spectrum is defined for an harmonic excitation. Depending on the propagation mechanisms involved, $\tilde{E}(k, v)$ can be a deterministic or a (partially) random function. Section 2.2.2 will only consider random plane-wave spectra.

Why should we be interested in random representations? Essentially because deterministic descriptions are possible but often unlikely or untreatable. Two reasons can be given about the need for random representations of field distributions in reverberation chambers: (1) their Green functions are highly complex and even numerical solutions are cumbersome, due to the resonant nature of these media; (2) reverberation chambers basically work as interferometers and therefore the field distributions they generate are very sensitive to the geometry and nature of their boundaries. In practice, the solutions sought for reverberation chambers do not imply a perfect knowledge of their fields at any position in a deterministic way, but more of a macroscopic evaluation of their performance. An example is given by the need to know the average field level a reverberation chamber can generate, with no need to know where (and even if) this specific value of field will appear. Hence the statistical description used for electromagnetic reverberation chambers [7,8] is predated by the work done in acoustics [9–12].

The reasons and motivations given above retrace those considered in the development of the kinetic theory of gases, pioneered by J. C. Maxwell and L. E. Boltzmann [13]. Coherently with this mindset, rather than looking for spatial solutions of fields generated by reverberation chambers (or any diffusive media), we rather need to start by arguing about the properties that should be expected for their plane-wave spectra and work our way back to understand spatial properties, as done in Section 2.2.2.

The assumption of perfect diffusion will be the cornerstone of this entire chapter. It requires that energy is randomly distributed along all possible directions, with equal probability and average intensity. The concept of probability is here defined across all possible realizations of field distributions (ensemble), not only the single distribution one could experience in a given configuration. Furthermore, for each direction, energy can be associated to any polarization for the electromagnetic field; e.g., this condition is used in optics to describe incoherent light.

Why should we expect these conditions to hold in practice? Fundamentally, there is no demonstration for their validity, as it would require the existence of exact solutions serving as references. The validity of these conditions are always validated a posteriori against experimental results. Qualitative reasons usually given for these assumptions are that the field radiated by a source within these media undergoes a number of scattering events that redistribute energy across an ever larger number of directions. Moreover, the fact that these chambers are electrically large implies that along any direction of propagation delay translates into uniformly random phase-shift angles. These two conditions are practically sufficient to invoke the concept of diffuse-field propagation.

From acoustics to electromagnetics, it appears to give satisfaction as soon as a large number of degrees of freedom are accessible at the same time. Without going into details, reverberation chambers need to be electrically large and present a certain degree of losses [2, 14].

If the field at any position is the result of a large number of random and independent plane waves, then the central limit theorem states that the field should asymptotically behave as a Gaussian random variable. The link between a random superposition of independent contributions and Gaussian theory goes back to the earliest works on Brownian motion [15].

Diffusion also has implications on the time-domain responses. A receiver within a diffusive medium would initially record only a few distinguishable echoes, which, as time passes and multiple scattering events take place, let an ever larger number of smaller and superposed echoes appear. The result is what is commonly known as reverberation, a phenomenon apparent at acoustical frequencies in large bounded environments like empty hangars and cathedrals. For our purposes, reverberation is important as it implies that by exciting a diffusive medium with a short pulse, its response can be several orders of magnitude longer, as discussed in Section 2.3.1. A pulse can be deformed in this way only if its frequency components are made incoherent, i.e., they appear to have both amplitude and phase-shift angles behaving as independent random variables. As an example, one of the reverberation chambers in our department responds with a time constant of about 3 μs, despite the fact that line-of-sight propagation through it just takes 20 ns, implying a very large number of interactions over its boundaries.

This condition leads to another fundamental property of diffusive media. The Wiener–Khinchin theorem implies that random-like impulse responses with energy distributions that fade very slowly are underpinned by narrow coherence bandwidths. In other words, the longer the decay time, the more incoherent is the frequency response, and thus richer in degrees of freedom that can be excited independently one from the other.

The broad picture evinced from these observations is that of a class of media that could hardly be expected to support coherent propagation of waves. As will be shown in the rest of this chapter, time reversal can completely revert this expectation, making diffusive media appealing for coherent applications.

2.2.2 Statistical Black-Box Modeling

The assumption of perfect diffusion can be translated into a number of macroscopic properties shared, at least approximately, by all diffusive media. The approximate nature of this approach comes from the impossibility to ensure an infinite number of degrees of freedom. Statistical moments are the language we need and they are applied across all possible random realizations by computing averages, indicated as $\langle \cdot \rangle$.

Perfect diffusion can be stated very succinctly using a plane-wave spectrum description:

$$\langle \tilde{E}^H(k_i) \cdot \tilde{E}(k_j) \rangle = \tilde{E}_o^2 \delta_{ij} \tag{2.2}$$

where $\tilde{E}(k_i)$ is the plane-wave spectrum of a field distribution, say the electric field, sampled at the wavenumber k_i. Equation (2.2) includes angular uniformity of the spectral power density $\tilde{E}_o^2$, uncorrelation of the individual plane-wave contributions, and depolarization.

The other assumption behind perfect diffusion is a uniform probability for all phase-shift angles of the plane waves. Their amplitudes just need to be random and although they can be expected to be zero-averaged, this condition has no impact on the corresponding spatial properties.

The first property that can be derived from these two conditions is the covariance of the spatial field. Defined as

$$C_E(r) = \langle E^H(r) \cdot E(r) \rangle = \frac{\langle |E(r)|^2 \rangle}{3} \mathbf{1} = \frac{E_o^2}{3} \mathbf{1} \qquad (2.3)$$

it results in a spatial field distribution whose scalar components are uncorrelated and share the same average intensity $E_o^2/3$, which also appears to be constant across space (homogeneity); $\mathbf{1}$ is the identity matrix.

When comparing field samples observed at different positions, correlations can be computed in order to evaluate whether the field samples are somehow coupled or not. This is the analog of assessing whether a time series has memory.

The spatial correlation coefficient, the normalized definition of spatial correlation, is found to be

$$\mu_r(||r_1 - r_2||, v) = \frac{\langle E^H(r_1, v) \cdot E(r_2, v) \rangle}{E_o^2} = \text{sinc}(k||r_1 - r_2||)$$

$$(2.4)$$

thus mostly concentrated in a region about half a wavelength wide, consistent with the qualitative observations in the speckle distribution in Figure 2.1. This is just one of the possible spatial correlation functions that can define the macroscopic behavior of diffusive media [16].

The frequency-domain behavior is mainly characterized by means of a correlation function

$$\mu_v(|v_1 - v_2|) = \langle E^H(r, v_1) \cdot E(r, v_2) \rangle / E_o^2 \qquad (2.5)$$

with a typical scale of correlation summarized by the coherence bandwidth

$$B_c = \int d\Delta v \mu_v(\Delta v) \qquad (2.6)$$

which is related to the average decay time, or relaxation time, τ of the medium as

$$\tau = 1/B_c \qquad (2.7)$$

Clearly, high-order statistics have been studied, but what has been presented in this section is the minimum set of properties needed in order to understand how time reversal works in diffusive media.

2.3 Time-Reversal Transmissions in Diffusive Media

Suppose a transmitter is used in a diffusive medium in order to reproduce a signal $p(t)$ at the receiver end, or equivalently a field at a given position, later also referred to as the target signal or pulse, with $P(v)$ its corresponding Fourier spectrum. If $H(v)$ is the transfer function between the transmitter and receiver, then directly applying $p(t)$ to the transmitter would result in the reception of an incoherent signal, since its spectrum $P(v)H(v)$ has lost any resemblance of the frequency coherence that characterized the spectrum of the originally intended signal $p(t)$. The reason behind this claim is to be found in the statistical properties of diffusive media, as recalled in Section 2.2.2: $H(v)$ has a short coherence bandwidth, outside which it behaves as an independent random function that will therefore act as random weights on $P(v)$. As a result, portions of $P(v)$ that were meant to be in phase and with slow amplitude variations will inherit the fast and random frequency variations of $H(v)$, yielding a received signal affected by long spread times.

Time reversal, in its original definition, provides a solution that is essentially equivalent to a matched filter, applying a signal $x(t)$ with a spectrum

$$X(v) = P(v)H^*(v) \tag{2.8}$$

to the transmitter, thus receiving a signal $y(t)$ with a spectrum

$$Y(v) = P(v)|H(v)|^2 \tag{2.9}$$

The Fourier spectrum in (2.8) corresponds to the time-reversed version of the signal received by the receiving antenna. Because of this operation, it is often referred to as a time-reversal mirror, or TRM.

Depending on the class of propagation medium, the received signal may have very different properties. Free-space propagation, and in general open media where line-of-sight propagation is a good approximation, implies that $|H(\nu)|^2$ is mostly flat or very slowly varying, so that it is reasonable to expect $Y(\nu) \propto P(\nu)$, i.e., a practically perfect transmission of the target signal $p(t)$ that was meant to be transmitted.

A very different behavior is observed in diffusive media. In Section 2.2.2 their frequency responses were shown to be affected by fast variations, with a scale of variation captured by their coherence bandwidth B_c. In the case where $P(\nu)$ spans a bandwidth $B_T \gg B_c$, (2.9) would be dominated by the variations in amplitude of the transfer function, profoundly modifying the original shape of $P(\nu)$. Equation (2.9) ideally removes the possibility of destructive interference, since phase shifts are equalized to zero for all frequencies. In effect, the original envelope on the target spectrum is hardly recognizable in $Y(\nu)$, as is clear in Figure 2.2.

However, time-reversal transmissions can be effective and lead to good transmissions of target signals at the receiver. The secret behind their effectiveness is qualitatively explained in Section 2.3.1 and later more demonstrations are given in Section 2.3.3.

2.3.1 Self-Averaging and Spectral Coherence

Self-averaging is probably the most impressive feature of time-reversal applications in diffusive media [17–19]. In the previous

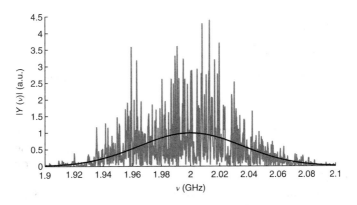

Figure 2.2 An example of the received spectrum $Y(\nu)$ (amplitude is shown in grey color) for a Gaussian target spectrum $P(\nu)$ (black line): the target Gaussian profile is hardly recognizable.

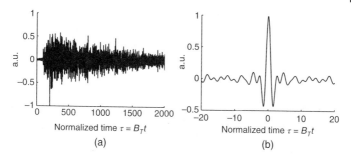

Figure 2.3 An example of impulse responses measured in one of CentraleSupelec's reverberation chambers, for $f_c = 1.1$ GHz and $B_T = 0.5$ GHz: (a) direct impulse response $h(t)$ and (b) equivalent impulse response $h(t) \star h(-t)$ for time-reversal transmissions. Time is normalized to the coherence time $1/B_T$.

section it was argued that the spectrum of signals transmitted with time reversal are so strongly modified by low-frequency coherence of the medium that it is surprising that back in the time domain these signals actually appear to be good replicas of a target signal.

Figure 2.3 provides a qualitative example of time-reversal performance in diffusive media. A medium with an impulse response thousands of times longer than the signal to transmit can generate a short pulse at the output of a receiver. These results only make to possible for the time compression to be enabled by time reversal; the spatial counterpart of this compression will be discussed in Section 2.4.

Several ways exist to expose and understand self-averaging. We propose to start from the time-domain representation of the received signal $y(t)$:

$$y(t) = p(t) \star h(-t) \star h(t) \tag{2.10}$$

with $h(t)$ the impulse response of the propagation medium and $\star$ the convolution operator applied to the time variable. The portion $y(-t) \star y(t)$ is a short step away from the autocorrelation function $R_{hh}(t)$ of the impulse response, defined as

$$R_{hh}(t) = \langle h(-t) \star h(t) \rangle \tag{2.11}$$

for $h(t)$ interpreted as a random process. It is worth relating the autocorrelation function of $h(t)$ to the received signal

$$\langle y(t) \rangle = p(t) \star R_{hh}(t) \tag{2.12}$$

as it has something to tell about the potential outcome of time-reversal transmissions.

In this last equation, the ensemble average of an *infinite* set of signals received in a given medium corresponds to the convolution of the target signal with $R_{hh}(t)$. Depending on whether $R_{hh}(t)$ distorts $p(t)$ or not, signals could be perfectly received, at least on average.

This would be a best case that is therefore worth studying. It can be done more easily in the frequency domain by recalling the Wiener–Khinchin theorem

$$R_{hh}(t) = \int_{B_T} d\nu \langle |H(\nu)|^2 \rangle e^{j2\pi t} \tag{2.13}$$

Assuming that the random process $H(\nu)$ follows a Gaussian law $N(0, \sigma_H(\nu))$, as reasonable for diffusive media (see Section 2.2), it is clear that in order to have $\langle y(t) \rangle \propto p(t)$, $\sigma_H(\nu)$ needs to be constant over the bandwidth B_T of $p(t)$; otherwise a distortion would appear, as $P(\nu)$ would then be weighted unequally.

Although this observation implies that time-reversal transmissions can potentially be perfect on average, in practice only one realization is available per transmitter. It is therefore fundamental to understand how close (2.10) is to (2.12) by studying

$$\Delta y(t) = y(t) - \langle y(t) \rangle = \int_{B_T} d\nu \, P(\nu)[W(\nu) - \langle W(\nu) \rangle] e^{j2\pi\nu t} \tag{2.14}$$

having introduced the time-reversal transfer function $W(\nu) = |H(\nu)|^2$. It is now clear that fluctuations in received signals are due to deviations of $W(\nu)$ from its average value. Hence, as a medium becomes more complex and multipath contributions start to interfere, its transfer functions can present local deviations from line-of-sight behavior, which is essentially flat in the frequency domain.

Going back to (2.14), the random process $W'(v) = W(v) - \langle W(v) \rangle$ is now centered, i.e., zero-averaged. Equation (2.14) can be written as

$$
\Delta y(t) = \sum_{n=1}^{N} \int_{B_n} dv P(v) W'(v) e^{j2\pi vt}
$$

$$
\simeq \sum_{n=1}^{N} P(v_n) \int_{B_n} dv W'(v) e^{j2\pi vt} \qquad (2.15)
$$

where the sub-bandwidths B_n are chosen to be identical over the entire bandwidth B_T, so that $N = B_T/B_n$. If B_n is small enough, $P(v)$ can be regarded as constant over B_n, and hence give the above approximation, where v_n is the central frequency of each sub-bandwidth B_n.

Equation (2.15) can be restated as

$$
\Delta y(t) \simeq \sum_{n=1}^{N} P(v_n) G_n \qquad (2.16)
$$

where G_n (implicitly defined) behave as random variables. If B_n is chosen such that the G_n are independent and identically distributed, (2.16) can be simplified as explained below. This condition depends on the integrands in (2.15), and therefore on the time t at which they are evaluated. For pulsed signals, most of their energy is found for $|t| < 1/B_T$. In this case, the phase shifts introduced by the Fourier kernel $e^{j2\pi vt}$ in (2.15) can be regarded as negligible over a number of adjacent sub-bandwidths. As a result, the G_n can be expected to be independent and identically distributed as soon as $B_n \geq B_c$. Since $H(v)$ fluctuates on a very short scale when compared to $P(v)$, this last condition is compatible with the idea of seeing $P(v)$ as constant over B_n.

Interpreting (2.16) as a weighted average of the G_n, i.e., of random variables, the central limit theorem can be invoked. Having chosen the B_n to ensure independence of the G_n, this theorem can be applied, concluding that the random variable $\Delta y(t)$ asymptotically follows a Gaussian distribution, centered around zero and with a standard deviation that scales as $1/\sqrt{B_T/B_c} = 1/\sqrt{N}$.

To make a long story short, time-reversed signals in diffusive media start to substantially converge to a target signal as soon as N exceeds 100. In diffusive media B_c can be very small; e.g., in a reverberation chamber operated at 1 GHz with a quality factor around 10^4, $B_c \simeq 300$ kHz, and hence a good convergence can be expected for $B_T > 30$ MHz, i.e., a fractional bandwidth of just 3%. This result is in sharp contrast with the bandwidths displayed in acoustics, where fractional bandwidths exceeding 50% are often used without a clear justification [20, 21].

More details about the performance of time-reversal transmission in diffusive media are presented in Sections 2.3.3 and 2.3.5, in particular about quantitative predictions of the accuracy of the received signals.

As a final remark about self-averaging in diffusive media, it is interesting to ponder the following question when looking at (2.12): what properties does $h(t)$ need in order for its autocorrelation function not to modify $p(t)$? Not modifying a band-limited function requires that $R_{hh}(t)$ be proportional to a cardinal sine (or sinc) function, with a Fourier spectrum spanning a frequency bandwidth containing that of $p(t)$, i.e., B_T. This condition is met in two extreme cases: a perfect channel that does not modify a signal or, paradoxically, one that has an impulse response behaving as a white noise, at least over B_T. In fact, diffusive media, as recalled in Section 2.2.2 do have impulse responses that resemble Gaussian random processes. It is therefore fitting to conclude that the most complex media that nature can produce can behave as the simplest one, once time-reversed signals are applied to them. This could also be seen as a further application of the Van Cittert–Zernicke theorem [22], where random processes can lead, under certain conditions, to deterministic outcomes.

2.3.2 Taking a Broader View: Spatial Speckle and Background Fluctuations

The previous section has highlighted a direct cause–effect link between rapid variations in transfer functions in diffusive media and random fluctuations in time-reversal transmissions, underpinned by a Fourier transform.

The same idea can be applied to another important couple of quantities also constituting a Fourier pair: spatial and angular field distributions. In fact time reversal is not only a way of compressing time signals but operates on both the space and time evolution of a wavefront as two facets of a single physical phenomenon. These two aspects should not be separated or regarded as independent.

While details about time-reversed wavefronts in diffusive media are given in Section 2.4, here we just want to give a preliminary discussion about why and how time-reversed wavefronts always appear as moving through a sort of noisy background. In the same way that the received and target signals are related by a convolution kernel, the spatial distribution of a time-reversed wavefront and the reference distribution moving through freespace are also related by a convolution kernel, even though it is more complex.

When this convolution is recast in its Fourier-transformed version, space is supplanted by directions of propagation of what could be interpreted as plane waves. Hence, as in the previous section, rapid variations in the propagation operator must lead to fluctuations in the spatial distribution of time-reversed fields.

Why should one expect rapid variations in the propagation operator? And what ultimately is this operator standing for? A spectral point of view has already been presented in Section 2.2 and basically models the fact that propagation within diffusive media, because of their boundary conditions, mostly happens along certain directions and, more importantly, with a varying intensity, due to constructive and destructive interference between multipath contributions.

Hence, there is no point in expecting time-reversed wavefronts to be free of background fluctuations, which typically take the shape of a speckle distribution, as defined in Section 2.2.1.

The intimate connection between spatial and time fluctuations can be evidenced by recalling that a receiving antenna fundamentally acts as a spatial (and partially angular) sampler [23]. In the simplest case, its output signals are just a copy of the fields found at its position. Therefore, in this simplified example, spatial fluctuations in the field distribution and time fluctuations in the received signal are the same. If the antenna has a marked directivity, than the two sets of fluctuations can differ, although their evolutions are still strongly related.

2.3.3 How Accurate are Time-Reversed Fields in Diffusive Media?

Signals, and also fields, generated by time reversal in diffusive media are not perfect; even worse, they have a random nature. Their accuracy can be measured, and ultimately predicted, by knowing just a few statistical metrics of the medium.

The first step is to define what is meant by accuracy. Since received signals are expected to get as close as possible to a target signals, it is possible to represent received signals as [24]

$$y(t) = \alpha\, p(t) + f(t) \tag{2.17}$$

i.e., as a partial perfect transmission of the target signal and a random process $f(t)$. In order to cover the most general scenarios, several transmitting antennas are supposed to take part in the transmission. Therefore, in general

$$y(t) = \sum_{m}^{N_A} \alpha_m p(t) + \sum_{m}^{N_A} f_m(t) \tag{2.18}$$

where N_A is the number of antennas. For each antenna a transfer function $H_m(\nu)$ can be defined between the transmitting and receiving antennas. The coefficients α_m can be computed by projections, e.g., in the frequency domain,

$$\alpha_m = E_p^{-1} \int_{B_T} d\nu\, W_m(\nu)|P(\nu)|^2 \tag{2.19}$$

with $W_m(\nu) = |H_m(\nu)|^2$ the time-reversal transfer function and E_p the mathematical energy of the target signal. Given that the functions in the integrand are positive defined, the coefficients $\alpha_m \geq 0$ by definition. Therefore multiple-antenna time-reversal transmissions cannot interfere destructively.

From (2.18) the accuracy of time reversal can be measured as the contrast between the coherent and incoherent contributions

$$\Lambda_p = \frac{\alpha^2 p^2(0)}{\max_t \langle f^2(t) \rangle} \tag{2.20}$$

having assumed, with no loss of generality, that the peak of the target signal is reached at $t = 0$. The above definition will be referred to as peak contrast as it is based on peak powers for the coherent (instantaneous power) and incoherent (average power).

The peak contrast can be factorized by introducing the shape factors

$$\chi_p = \frac{|p(0)|^2}{E_p}$$

$$\chi_f = \frac{\max_t \langle f^2(t) \rangle}{\langle E_f \rangle}$$

(2.21)

which measure the coherence bandwidth of the target signal and incoherent fluctuations; (2.20) can now be expressed as

$$\Lambda_p = \Lambda \frac{\chi_p}{\chi_f}$$

(2.22)

with

$$\Lambda = \frac{E_c}{E_f} = \frac{\alpha^2 E_p}{\int_{B_T} d\nu |F(\nu)|^2}$$

(2.23)

the energy contrast, which now measures the ratio of the coherent and incoherent energies. The rationale behind this factorization is that it is possible to prove [25] that Λ only depends on the statistical properties of the medium, and not those of the target signal. The derivation of an explicit relationship between Λ and the medium statistical properties is cumbersome and therefore cannot be reproduced here. The result derived in [25] proves that

$$\langle \Lambda \rangle \simeq \varsigma_W^{-2}(\nu_c) D(N_A, \bar{\mu}_r)$$

(2.24)

with

$$\varsigma_W^2(\nu) = \frac{\sigma_W^2}{\langle W(\nu) \rangle^2}$$

(2.25)

the statistical variability of the time-reversal transfer function, i.e., whose square-root value measures how strongly $W(\nu)$ can deviate from its average value. For perfectly diffuse fields, it is equal to 1. The other term in (2.24) is the diversity factor,

$$D(N_A, \bar{\mu}_r) = \frac{N_A}{1 + (N_A - 1)\bar{\mu}_r} \tag{2.26}$$

which measures the effective number of degrees of freedom introduced by the N_A transmission antennas, for an average correlation coefficient $\bar{\mu}_r$ between the $W_m(v)$ time-reversal transfer functions.

Two conclusions can be drawn. First, the medium itself sets a scale of the performance of time-reversal transmissions. While the peak contrast Λ_p can be controlled by acting on the bandwidth of the excitation signal, the fraction of received energy associated with the target signal is definitely fixed by the medium.

Second, having several antennas cooperating with the transmission can help, since the diversity factor $D(N_A, \bar{\mu}_r) \geq 1$. Therefore, using time-reversal mirrors in diffusive media still makes sense, even though not for the same reasons as in open media, i.e., not in order to define a time-reversal cavity (e.g., a Huygens' surface). However, depending on the independence of the time-reversal transfer functions $W_m(v)$, adding antennas could represent an ineffective solution. $D(N_A, \bar{\mu}_r)$ is plotted in Figure 2.4, where it is clear that as soon as a small average correlation appears, the increase in the energy coherence is

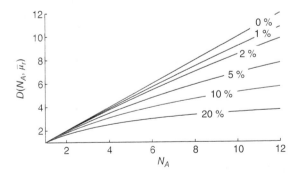

Figure 2.4 The diversity factor $D(N_A, \bar{\mu}_r)$ as a function of the number of sources N_A and the average spatial degree of coherence $\bar{\mu}_r$ in percent units over each curve.

reduced. This reduction is stronger as the number of antennas already present increases. Since correlations around 20% are all but rare, multiple-source configurations can be costly, as they risk bringing very minor improvements while forcing the need for synchronized excitations. This last point has practical importance, since synchronized generation of multiple wideband signals is a far from obvious task.

As just recalled, while Λ can be difficult and costly to control, (2.22) states that the peak contrast does not only depend on the medium through χ_f but can also be modulated by accurately designing the target signal acting on χ_p. Equation (2.22) can be made clearer by noticing that the two shape factors can be expressed in terms of characteristic time constants: e.g., $\chi_f \propto 1/T_f$, with T_f the relaxation time of the medium, i.e., the time constant of its root-mean-square impulse response; in a similar way, $\chi_p \propto 1/T_p$, with T_p the time support of the target pulse, e.g., defined as the portion of pulse containing half of its energy or where the pulse passes to half of its peak power. Hence making the target signal shorter makes the peak contrast increase.

The different impacts of pulse design and number of antennas are experimentally demonstrated in Figure 2.5, where signals received with different configurations are compared. These results confirm that increasing the number of antennas, even when weakly correlated ($\bar{\mu}_r \simeq 10\%$), can be very ineffective, whereas bandwidth is a safe and usually cheaper solution.

While contrast is fundamental in order to have a proper transmission, the term accuracy also has another meaning: to have a stable transmission, i.e., to know beforehand the absolute level of the received signals, rather than just their relative value with respect to a random background. The derivation of (2.16) led to the conclusion that time-reversed transmissions generate signals (and fields) random in amplitude, behaving as Gaussian distributed random variables with standard deviations scaling as $\sqrt{B_c/B_T}$. This prediction was experimentally validated in [26], for target signals centered at 2 GHz and a varying bandwidth. Figure 2.6 compares empirical histograms with the theoretical probability density function of a standardized Gaussian random variable. Even in the case of $B_T = 2$ MHz, i.e., for $B_T/B_c \simeq 4$, the probability distribution of the amplitude of the received signals is consistent with the theoretical one.

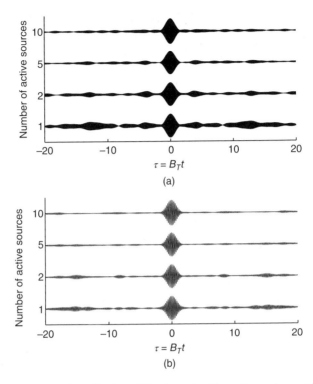

Figure 2.5 A single realization of signals received for an increasing number of active sources, for $v_c = 0.7$ GHz and: (a) $B_T/v_c = 5\%$; (b) $B_T/v_c = 20\%$. The peak values attained by the signals are practically independent of the number of sources.

This agreement implies that given the maximum admissible variations in received signals, the minimum bandwidth B_T of the target signal can be chosen in a straightforward manner.

2.3.4 Polarization Selectivity

Time reversal is usually demonstrated with scalar quantities. Having originated in acoustics, this is not surprising, but applications to electromagnetics have also been mainly limited to received signals, and thus are intrinsically scalar [19, 27–32].

However, time reversal used in the case of vector fields is of interest. When done in diffusive media this takes on a particular

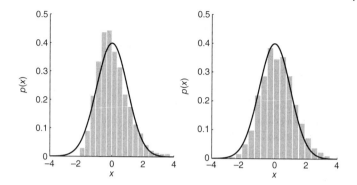

Figure 2.6 Empirical and theoretical pdfs of the standardized variable $(||y(t)||_\infty - u)/\sqrt{u}$, with $u = \bar{W} B_T / B_c$, for $B_T = 2$ MHz (left) and 32 MHz (right). These results confirm the statistical convergence of the peak of time-reversed signals, as controlled by u. $\bar{W}$ is the average value of the time-reversal transfer function $W(v)$ computed over B_T.

significance and leads to another remarkable property: a single scalar signal can control the polarization of a vector field, just by changing its time evolution. This property was first demonstrated in [33].

In order to understand how a scalar excitation could control vector fields, let us consider the vector transfer function $\mathbf{\Phi}(v)$ between a voltage $X(v)$ applied to the input port of an antenna and the electric field it generates at a given position within a diffusive medium, such that

$$E(\mathbf{r}, v) = \mathbf{\Phi}(v)X(v) \tag{2.27}$$

where the transfer function can be expanded into a (eventually infinite) summation of eigenmodes $e_m(\mathbf{r}, v)$, i.e., solutions to the sourceless version of the Helmholtz equation, as

$$\mathbf{\Phi}(v) = \sum_m^M \gamma_m e_m(\mathbf{r}, v) \tag{2.28}$$

with $\mathbf{r}$ the position at which the field is sampled; the γ_m are modal coefficients expressing the rate of excitation and relative phase shift of the modes.

In disordered media, and particularly in diffusive media, fields are expected to be not only randomly distributed but also randomly polarized, as recalled in Section 2.2, i.e.,

$$\langle \boldsymbol{\Phi} \, \boldsymbol{\Phi}^H \rangle \propto \mathbf{1} \tag{2.29}$$

with $\mathbf{1}$ the identity matrix and H the Hermitian transpose. Equation (2.29) contains multiple statements: not only does each polarization have the same average intensity but also each couple of different polarizations are ideally uncorrelated. This property is expected for each frequency only across multiple random realizations.

In practice, the idea of having to generate a set of realizations in order to access an interesting property is a big turn-off, making a potentially useful property more of a theoretical observation.

Time-reversed signals work across the frequency axis and can be shown to make this property accessible and usable in practice. It has been observed in [34] that exciting a reverberation chamber at different frequencies is equivalent to generating a set of random realizations. In spite of the fact that this property evokes the idea of ergodicity, it would be incorrect to take it for granted. A better explanation of this property is to be sought in the statistical properties of resonant frequencies of large cavities, as studied in wave chaos theory [35]. We will leave this topic to the interested reader and focus on this apparent ergodicity for our practical purposes.

In short, as the modes M in a medium excited over the bandwidth B_T of a target signal increases, the number of modes contributing to the overall field will also increase, averaging out their random contributions. If the random properties of these modes are assumed (as discussed above) to be identical when looking at different realizations or at different frequencies, the following asymptotic equivalence holds:

$$\lim_{M \to \infty} \int_{B_T} \mathrm{d}\nu \, \boldsymbol{\Phi}(\nu) \, \boldsymbol{\Phi}^H(\nu) \propto M \langle \boldsymbol{\Phi}(\nu_c) \boldsymbol{\Phi}^H(\nu_c) \rangle \tag{2.30}$$

This abstract-looking result is in fact of great practical utility. Let us now consider an excitation signal

$$X(\nu) = P(\nu) \boldsymbol{\Phi}^H(\nu) \hat{\boldsymbol{p}} \tag{2.31}$$

where $P(\nu)$ is the Fourier spectrum of the target signal and $\hat{p}$ is the polarization along which this signal should appear as a field. In case the target signal is a pulse reaching its peak around $t = 0$, the resulting field reads

$$E(r, 0) = \int_{B_T} d\nu P(\nu)\mathbf{\Phi}(\nu)\mathbf{\Phi}^H(\nu)\hat{p} \tag{2.32}$$

which, in the case of an ideal diffusive medium, simplifies thanks to (2.29) and (2.30) to

$$E(r, 0) \simeq E_o\hat{p} \tag{2.33}$$

where E_o is the root-mean-square intensity of any of the scalar components of $\mathbf{\Phi}(\nu)$, i.e., directly related to the energy density excited within the medium. The trick behind this result is the same already used to understand self-averaging in Section 2.3.1, i.e., taking advantage of the different frequency scales over which to operate the medium transfer functions and the target spectrum.

Equation (2.33) states that asymptotically time-reversed fields can exert perfect control of their polarization for any chosen target polarization $\hat{p}$.

This theoretical prediction is proven to be correct in Figure 2.7, where the Cartesian field components generated by time-reversal excitations as in (2.31) are shown; three different choices for the final field polarization $\hat{p}$ are considered. Random background fluctuations are still present, as shown in Section 2.3.2, so field polarizations can be controlled only when the target signal reaches its strongest intensity.

Figure 2.7 shows just one realization of time-reversed fields. Repeating these experiments over 50 positions distributed in a reverberation chamber led to the results shown in Figure 2.8 [36].

These results prove that in a diffusive medium, well-polarized fields can be generated even with antennas not intended for this purpose. Polarization rejections better than 25 dB were observed in half of all the cases tested; in 90% of the experiments the rejection was close to 20 dB. A generic rejection is indicated as $\rho_{ij} = E_i/E_j$, with E_i one of the two cross-polarization components and E_j the co-polarization one.

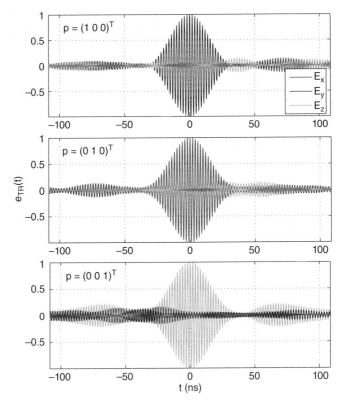

Figure 2.7 Field components obtained from experimental results measured at one position, for the Gaussian pulse at 1.5 GHz. Each plot refers to weight vector **p** corresponding to one Cartesian direction. Top to bottom, x, y, z components of fields are ideally only excited when the pulse attains peak value.

These results laid the foundations for the generation of coherent wavefronts within reverberation chambers, presented in Section 2.4. Indeed, only when polarizations can be controlled can a wavefront be generated. Being able to do so with a single antenna bearing no specific features makes it possible to define novel applications for time reversal, but only when applied to diffusive media.

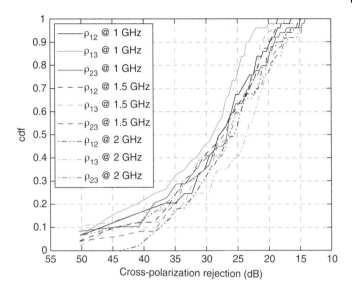

Figure 2.8 Empirical probability distribution for cross-polarization rejections experimentally observed in time-reversed fields, generated to have a dominant linear polarization.

2.3.5   Conversion Efficiency and Power Gain

The space–time compression operated by diffusive media excited with time-reversed excitation can be exploited in a further way. Suppose an input signal is applied to an antenna used for exciting a reverberation chamber. After a transient, the energy injected is on average uniformly distributed over the entire volume of the chamber. A receiver would thus intercept at a given time only a small portion of this energy, which will meanwhile keep propagating *incoherently* throughout the volume of the chamber. If the receiver were an electric equipment tested within the chamber, then it would be submitted to a relatively weak electromagnetic stress with respect to the total energy present inside the chamber.

In the case of time reversal, this picture can be partially subverted by having at least a portion of the energy converging (or focusing) *coherently* on to the equipment under test, at a given position and time. However, at the same time, time-reversed signals generated in a reverberation chamber have an extremely low peak-to-average ratio, as shown in Figure 2.3a. In other words, for a given input energy, the peak power attained by the excitation signal can be a fraction of the signal that will be focusing within the chamber.

This idea was first suggested in [37] and can be formalized by introducing the notion of conversion efficiency as the ratio between the peak instantaneous power observed by a receiver and the one from the input signal:

$$\eta_S = \eta[x_S(t)] = \frac{||p_r(t)||_\infty}{||p_i(t)||_\infty} \tag{2.34}$$

where the received power is defined with respect to an ideal sampler of the $\hat{q}$-aligned field component

$$p_r(t) = C_r[e(t) \cdot \hat{q}]^2 = C_r f^2(t)\hat{q} \tag{2.35}$$

with C_r a constant modeling the antenna factor of the sampler, e.g., a probe. The definition of the input power refers to the excitation signal $x(t)$,

$$p_i(t) = C_i x^2(t) \tag{2.36}$$

The measured field $f(t)$ is related to the excitation signal by means of a transfer function $\Phi(v)$, such that $f(t) = F^{-1}\{\Phi(v)X(v)\}$, with F^{-1} the inverse Fourier transform and $X(v)$ the Fourier spectrum of $x(t)$.

The conversion efficiency can then be computed for a continuous-wave (CW) excitation of a reverberation chamber as

$$\eta_{CW} = \frac{C_r}{C_i} W(v_c) \tag{2.37}$$

where $W(v) = |\Phi(v)|^2$ is the time-reversal transfer function introduced in Section 2.3.1.

Computing the same quantity for time-reversed excitations is trickier. In this case, $x_{TR}(t) = p(t) \star \phi(-t)$, where $p(t)$ is the target pulse meant to be reproduced at the probe position. One of the difficulties is to take into account the fact that the peak power of a time-reversed excitation is now a random quantity. This problem is addressed in [26] by introducing a multiplier K

modeling how much $\phi(t)$ can exceed its root-mean-square envelope, corresponding to $\langle|\phi(t)|^2\rangle = A_o^2 \exp(-2t/\tau)$, thus yielding

$$K^2 = \|p_{i(t)}\|_\infty / A_o^2 \tag{2.38}$$

The K factor, dubbed the overshoot factor, is a random quantity; the interested reader can find a derivation of its probability distribution in [26].

The main impact of the randomness of the time-reversed excitation is the impossibility to ensure a fixed peak power. As a result, for a given output field, the excitation required could vary, thus displaying a random efficiency. This is the opposite of what happens for the CW efficiency: in that case the input excitation is deterministic and the output field is random. The main difference between the two cases is the probability distribution behind these quantities, which has a very different nature, as discussed in [26]; the dispersion is much higher for CW excitations.

The resulting efficiency is given by

$$\eta_{TR} = \frac{C_r}{C_i} \frac{B_T}{B_c} \frac{\kappa^2}{K^2} \bar{W}(v_c) \tag{2.39}$$

where the overbar stands for the average value computed over the bandwidth B_T of the target pulse. The shape factor κ^2 measures how pulse-like the target signal is, and is defined as

$$\kappa^2 = B_T \bar{P}^2 / E_p \leq 1 \tag{2.40}$$

where $\bar{P}$ is the average of the spectrum of the target signal and E_p its mathematical energy.

Comparing (2.37) and (2.39) two differences come to mind. First, the factor B_T/B_c in the time-reversal case can be expected to lead to much higher efficiencies; its physical origin is to be sought in the self-averaging mechanism offered by diffusive media, summarized in Section 2.3.1, and is the main reason for the improved efficiency of time-reversed excitations. It was already pointed out in Sections 2.3.1 and 2.3.3 as the main asset of time reversal in diffusive media.

Second, while the CW efficiency is based on a single realization of $W(v)$, thus wildly random, the time-reversal one is rather based on its average $\bar{W}(v_c)$, which can be made practically deterministic. We have already pointed out the fact that the overshoot factor in (2.39) is also random but it is actually much less dispersed than $W(v)$, which follows an exponential probability function, i.e., with a standard deviation equal to its average, and a single realization as in (2.37) is therefore a risky bet. Hence the need for CW excitations for stirring solutions, in order to generate multiple independent realizations and to increase the chances of producing higher efficiencies.

Indeed, a more realistic definition of CW efficiency should take into account N runs at generating high fields from the same input signal, equivalent to throwing a dice several times and keeping the best outcome, for instance, as in

$$\eta_{CW}^N = \max_{i \in [1,N]} \overset{\bullet}{\eta}_{CW}^{(i)} \tag{2.41}$$

This extended definition is still badly dispersed, but the average of the CW efficiency can now be increased, though ineffectively, by increasing N.

In order to compare the two efficiencies their respective statistical models are taken as good representations of typical values. The power gain

$$G_P^N = \frac{\text{Mo}[\eta_{TR}]}{\text{Mo}\left[\eta_{CW}^N\right]} \tag{2.42}$$

is now regarded as a potential gain in the conversion efficiency enabled by time-reversal excitations. The reason for calling it a power gain is to be understood by noticing that, for a fixed output power, an increase in the conversion efficiency would correspond to a reduction in the input power needed.

It can be proven that (2.42) is well approximated by [26]

$$G_P^N = \frac{B_T}{B_c} \frac{\kappa^2}{\ln N} [a \ln^b (4B_T/B_c)]^{-2} \tag{2.43}$$

with $a = 0.749$ and $b = 0.678$. The power gain is now clearly dominated by the $B_T/B_c \ln N$ term, which exposes the inefficiency of the CW approach. Having N degrees of freedom (perfect stirring states) just increases the efficiency as their natural logarithm, while time-reversed excitations can take full advantage of B_T/B_c degrees of freedom. This difference is due to the

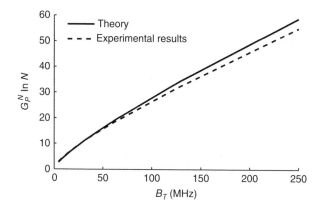

Figure 2.9 Normalized gain $G_p^N \ln N$ as a function of B_T.

collaborative nature of time reversal, where the B_T/B_c degrees of freedom are coherently excited. This result was tested in [26] against experimental results, finding a good agreement with theory, as reported in Figure 2.9.

The differences in the efficiencies can be better grasped by imposing an equal typical performance for the two excitations, so that (2.43) can be converted into

$$ N = \exp\left\{ \frac{B_T}{B_c} [a \ln^b (4 B_T/B_c)]^{-2} \right\} \tag{2.44} $$

for $\kappa = 1$. This last result estimates the typical number of stirring states needed for a CW excitation to compare favorably with time-reversal excitations, and is shown in Figure 2.10.

Not only are time-reversed excitations more efficient thanks to self-averaging but are also less statistically dispersed, since coherent excitation of the degrees of freedom is equivalent to a sample average. The central limit theorem applies in this case, implying a convergence to the average. Figure 2.11 shows the 95% confidence intervals computed for the two excitation schemes normalized to their respective average values as the number of available degrees of freedom is increased. The steady convergence of time-reversed excitations leads to fields characterized by amplitudes that can be practically considered as

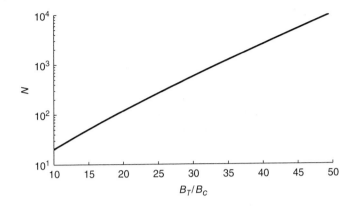

Figure 2.10 Equivalent number of independent realizations required in a CW-driven reverberation chamber, in order to ensure the same typical efficiency as when time-reversal driven, as defined by (2.44).

deterministic with a hundred degrees of freedom, whereas CW excitations would still generate typical relative dispersions of about 100%.

These results point to the possible use of time-reversed excitations as a threefold solution: stronger fields, more predictable, and in one shot, i.e., without repeating tests along stirring positions.

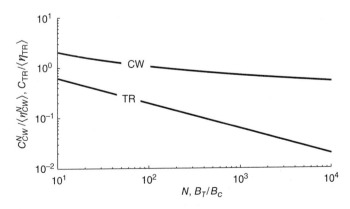

Figure 2.11 Relative confidence intervals of the conversion efficiencies for a 95% probability as a function of the number of degrees of freedom.

2.4 Time Reversal for the Generation of Wavefronts

So far time reversal has been applied to diffusive media in order to produce either an output signal at a receiving antenna or as a way of generating a field at a specific position. Both cases can be seen as a point-to-point transmission scheme.

This section rather explores the use of time reversal as a way of generating wavefronts within a diffusive medium. This possibility was first empirically demonstrated in a number of papers [20, 38–40], with the first demonstrations in acoustics. The question of how well a time-reversed wavefront is reproduced in a diffusive medium is first addressed from a theoretical point of view in Section 2.4.1, by introducing the time-reversal dyadic functions that relate the wavefront generated by means of time reversal within a given medium to the target wavefront expected to be generated.

The first phase usually taken for granted in time-reversal applications is argued in Section 2.4.2 to be a major obstacle in the development of test facilities based on time reversal. From these observations, a generalized approach to time reversal is presented in Section 2.4.3, where the first phase is avoided by directly synthesizing the excitations signals that would usually be expected to be recorded during the first phase.

A practical implementation of generalized time reversal is then discussed in Section 2.4.4, describing a robot prototype developed in CentraleSupélec. Experimental results obtained with it are shown and analyzed in Section 2.4.5, validating the possibility of designing test facilities based on time-reversal principles.

2.4.1 Time-Reversal Dyadic Function

With reference to Figure 2.12, consider a source of radiation occupying a region Ξ. Modeling it as a combination of electric and magnetic currents, respectively noted as $J_e^s(r)$ and $J_m^s(r)$, the electric field it produces can be written as

$$E_{wf}(r) = \int_\Xi dr' G_{ee}(r,r') \cdot J_e^s(r') + \int_\Xi dr' G_{em}(r,r') \cdot J_m^s(r')$$

(2.45)

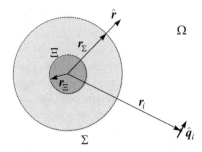

Figure 2.12 Configuration for the application of Love's equivalence theorem. Equivalent electric and magnetic currents are defined over the surface, representing the wavefront $E_{\mathrm{wf}}(r, t)$ that would have been generated by a synthetic source contained in the volume bounded by the surface Σ. The dipole polarized along $\hat{q}_i$ is a TRM antenna. These elements are embedded within a diffusive medium.

where the integrals are computed over the volume of Ξ and $G_{ee}(r, r')$ and $G_{em}(r, r')$ are the Green functions of the propagation medium for the electric field generated by electric and magnetic currents, respectively. In order to derive the relationship between the wavefront generated by means of time reversal and that radiated by the original source, it is convenient to define an auxiliary surface Σ, chosen spherical of radius r_Σ, as in Figure 2.12; this surface can be in the near-field region of the source.

Sampling the electromagnetic field radiated by the source over Σ, equivalent electric and magnetic currents can be defined, as stated by the equivalence theorem [41]:

$$
\begin{aligned}
J_e(r) &= -\zeta_o^{-1}\delta(r - r_\Sigma)\mathbf{1}_t \cdot E_{\mathrm{wf}}(r) \\
J_m(r) &= \delta(r - r_\Sigma)E_{\mathrm{wf}}(r) \times \hat{r}
\end{aligned}
\tag{2.46}
$$

where $\delta(\cdot)$ is the Dirac distribution, here used to represent sampling over the surface Σ; ζ_o is the wave impedance of the background medium, e.g. air; and $\mathbf{1}_t$ is the transverse identity dyad, which strips a vector of its radial component, leaving only the transversal components intact, here used in order to extract the tangential components of the electric field with respect to the surface Σ.

The field radiated by the source can now be written as

$$E_{wf}(r) = \int_{\Sigma} dr' G_{ee}(r, r') \cdot J_e(r') + \int_{\Sigma} dr' G_{em}(r, r') \cdot J_m(r')$$

(2.47)

where integrals are now taken over the surface Σ and involve the equivalent current distributions. If the TRM antenna in Figure 2.12 has an effective height $h_e(\nu) = h_e(\nu)\hat{q}$, eventually frequency dependent, the following vector transfer functions can be introduced:

$$N_e(r) = G_{ee}(r, r') \cdot \hat{q}$$
$$N_m(r) = G_{em}(r, r') \cdot \hat{q}$$

(2.48)

so that the signal received by the TRM antenna during the first phase reads as

$$V(\nu) = C_1(\nu) \left[\int_{\Sigma} dr' N_e(r') \cdot J_e(r') + \int_{\Sigma} dr' N_m(r') \cdot J_m(r') \right]$$

(2.49)

with $C_1(\nu)$ a scalar taking into account all frequency-related effects, such as an eventual dispersion of the receiving antenna.

Injecting the phase-conjugated (or time-reversed) version of $V(\nu)$ into the antenna, i.e., $V_{in}(\nu) = V^*(\nu)$, the field thus generated is given by

$$E_{TR}(r) = C_2(\nu) \left[\zeta_0^{-1} \int_{\Sigma} dr' N_e^*(r') N_e(r) \cdot E_{wf}^*(r') \right.$$
$$\left. - \int_{\Sigma} dr' \hat{r} \times N_m^*(r') N_e(r) \cdot E_{wf}^*(r') \right]$$

(2.50)

with $C_2(\nu)$ now also taking into account the frequency dependencies during the transmission phase.

Fields propagating through diffusive media can be treated as random variables, as discussed in Section 2.2, so that $E_{TR}(r, \nu)$ should at first sight also be regarded as random in nature. While technically correct, the randomness of $E_{TR}(r, \nu)$ can be practically ignored, as long as the wavefront $E_{wf}(r, \nu)$ covers a bandwidth at least two orders of magnitude wider than the coherence bandwidth of the medium. Under these conditions, the self-averaging property discussed in Section 2.3.1 is very effective, so that $E_{\mathrm{wf}}(r, v) \simeq \langle E_{\mathrm{wf}}(r, v) \rangle$, whence

$$
E_{TR}(r) \simeq C_2(v) \left[\zeta_o^{-1} \int_\Sigma dr' \, T_{ee}(r, r') \cdot E_{\mathrm{wf}}^*(r') \right.
$$

$$
\left. - \int_\Sigma dr' \, T_{em}(r, r') \cdot E_{\mathrm{wf}}^*(r') \right] \tag{2.51}
$$

where

$$
T_{ee}(r, r') = \langle N_e^*(r') N_e(r) \rangle
$$
$$
T_{em}(r, r') = \langle \hat{r} \times N_m^*(r') N_e(r) \rangle \tag{2.52}
$$

are dyadic functions dependent on the spatial correlation functions of the medium. The wavefront generated by means of time reversal can therefore be expressed as

$$
E_{TR}(r) \simeq C_2(v) \int_\Sigma dr' \, T(r, r') \cdot E_{\mathrm{wf}}^*(r') \tag{2.53}
$$

with

$$
T(r, r') = \zeta_o^{-1} T_{ee}(r, r') - T_{em}(r, r') \tag{2.54}
$$

the time-reversal dyadic function. Equation (2.53) implies that wavefronts generated by means of time reversal are not mere replicas of an originally radiated wavefront $E_{\mathrm{wf}}(r, v)$, but rather appear to be subject to a point-spread function, $T(r, r')$, in the shape of a dyadic function. This function may be expected to distort the original wavefront, both in its spatial evolution and polarization.

The previous results are completely general and can be applied to any medium, be it modeled as deterministic or random. For the special case of diffusive media, Section 2.2.2 recalled that the field generated there has statistical properties that are invariant with respect to rotation and translation, but not polarization. These properties were used in the derivation presented in [42], deriving the time-reversal dyadic function for the case of diffusive media. Although the derivation is not discussed here, Figure 2.13 gives an example of the point-spread function, when sampled over a spherical surface. These data fundamentally show the distribution of the electric field that would be generated

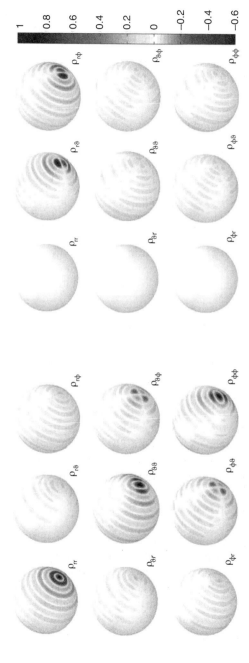

Figure 2.13 Normalized dyadic function $\bar{\bar{\rho}}(\boldsymbol{r}, \boldsymbol{r}', \nu)$ computed for $\boldsymbol{r} \in \Sigma$ and $\boldsymbol{r}' = r_\Sigma \hat{\boldsymbol{x}}$, with $r_\Sigma = 3\lambda$: (a) real and (b) imaginary parts. The nine terms of the dyadic response are shown, matrix-wise, considering standard spherical unit vectors, following the order $\hat{\boldsymbol{r}}$, $\hat{\boldsymbol{\vartheta}}$, and $\hat{\boldsymbol{\varphi}}$, defined with respect to a polar axis vertically oriented.

(on average) by time reversal, when attempting to reproduce a singular field distribution, i.e., a spatial pulse. Rather than reproducing the same distribution, Figure 2.13 shows that spatial resolution is limited and that, depending on the polarization of the original field, that of the time-reversed wavefront can be critically affected.

The point-spread function basically appears for two reasons. First, time reversal operates on signals received by antenna typically electrically distant from the original source. Hence, mostly propagative components are sampled, while reactive ones, those associated to near-field characteristics, are lost. Second, focusing purely propagative components cannot reproduce the polarizations found in the reactive region of the source, as attested by the cross-terms in Figure 2.13. In other words, as long as only the propagative part of a wavefront is needed, this can be reproduced accurately by time reversal, even in a diffusive medium. Otherwise, solutions such as time-reversal sinks would be required [43] in order to reintroduce the lost reactive contributions.

2.4.2 The Burden of a First Radiating Phase

Section 2.3 recalled how time-reversed signals are generated. In particular, it made clear that the very expression "time reversal" is in fact a truncated sentence, as it lacks an object. Indeed, it is not time that is reversed, but rather the time evolution of a wavefront; in particular, a wavefront first generated, e.g., by a source, like an antenna. Only after a TRM antenna has recorded the signature of the wavefront and the distortion introduced by the propagation medium can time reversal operate its magic.

Figure 2.14 summarizes the two existing approaches used in time reversal in open media. The first method corresponds to what has just been recalled: a source of radiation is activated generating a diverging wavefront; after recording by the TRM antennas, the propagative part of the wavefront can be generated, this time as a wavefront converging back to its origin.

While impressive in its effectiveness, this approach has a clear drawback: it is not possible to generate any wavefront on command, but only to reproduce what has been generated first by a source of radiation. Any application of time reversal aiming at

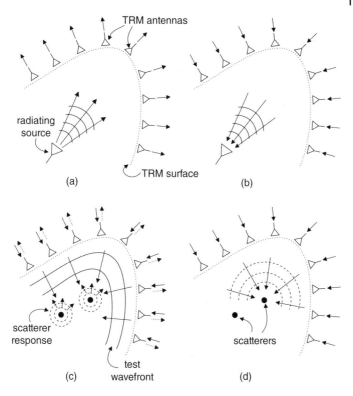

Figure 2.14 A schematic representation of the two main time-reversal techniques currently available: (a) and (b) time reversal of a radiating source; (c) and (d) selecting focusing over a point scatterer by means of the DORT approach.

generating focusing wavefronts in complex media is therefore hampered by the need: (1) to have an appropriate source of wavefronts that would then be time reversed; (2) to repeat this first phase of radiation before being capable of generating the actual wavefront of interest, the focusing one. For imaging the case of a set of wavefronts to be generated, with different features, e.g., direction of arrival and directivity, an equivalent set of sources with different orientations will need to be available (and characterized). Such a cumbersome procedure goes against the very idea of a test facility, which is expected to be capable of generating test wavefronts straight away.

Suppose now the case of a passive piece of equipment, to be submitted to an impinging focusing wavefront. Following the above description, the diverging version of the wavefront will need to be generated, but by what source? The equipment under test is passive. Should an antenna, acting as an external source, be added over the position where the focusing wavefront will converge? But how to ensure that the field radiated by the antenna would not be modified by the boundary conditions imposed by the equipment?

A solution to this problem comes with the second configuration described in Figure 2.14. In this case, rather than time-reversing the wavefront radiated by a source, it is rather the wavefront scattered by the equipment that is focused back at its origin. This idea gave way to the DORT (Décomposition de l'Opérateur de Retournement Temporel) method [44,45], where an object under test is submitted to a wavefront sounding it. The field thus scattered is recorded, post-processed through an eigenanalysis in order to define wavefronts that will focus over the brightest scattering points. Despite providing a solution to extend time reversal to the case of passive equipments, wavefronts thus generated are totally dependent on the equipment itself; necessarily only wavefronts corresponding to the equipment scattering can be reproduced. Moreover, the first phase is still there.

The rest of this section presents a solution capable of generating time-reversed wavefronts without involving any initial source. The method, denoted as generalized time reversal, or GTR, is first theoretically derived; requirements and limitations are then discussed; experimental results showing its ability to generate arbitrary wavefronts in a reverberation chamber are finally shown.

2.4.3 Direct Wavefront Synthesis

The derivation of the time-reversal dyadic function comes with an important byproduct. Equation (2.49) can be restated as

$$V(v) = C_1(v) \int_\Sigma d\mathbf{r}' \mathbf{N}_{eq}(\mathbf{r}') \cdot \mathbf{E}_{wf}(\mathbf{r}') \qquad (2.55)$$

where

$$N_{eq}(r) = -\zeta_o^{-1} N_e(r) + \hat{r} \times N_m(r) \tag{2.56}$$

Equation (2.55) now states that the voltage that would be received in the presence of a source can actually be computed, as soon as the medium Green's function between the TRM antenna and the surface Σ are known and converted into the function $N_{eq}(r)$.

The physical meaning of (2.55) is schematically represented in Figure 2.15, where the surface Σ acts as an interface between the field radiated by a source (step 1) and the TRM antenna (step 2). By performing step 1 offline, e.g., by using a closed-form expression for the field radiated by a source, or a numerical model of the source, (2.56) yields the received voltage. In the end, this process is equivalent to a virtual first phase, where sources of any kind are available. That said, it should be clear that the quality of time-reversed wavefronts strongly depends on their space and time evolution. A discussion about this point is presented in Section 2.4.5.

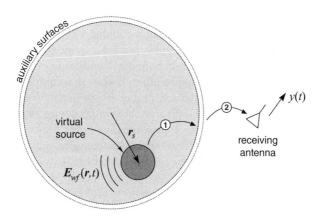

Figure 2.15 Relevant quantities in the derivation of direct synthesis of excitation signals with GTR. The auxiliary surfaces are shown to act as an interface between the virtual source radiation (1) and its propagation through the medium (2).

This result has practical implications, since it has the potential of bypassing the need for a first phase in time-reversal applications; the wavefront that would be generated by any source could be time reversed just by describing the field distribution it would generate over Σ. Since the entire procedure represents the propagation through the medium up to the TRM antenna, only propagative contributions will survive. For this reason, even if Σ were in the near-field region of the virtual source, only its propagative components need to be sampled over Σ. Therefore, the far-field radiation of the virtual source is sufficient, which is also simpler to compute than a full description of radiation valid also in the near-field region.

Direct synthesis is not the only advantage of this procedure. As discussed in Section 2.3.3, a single antenna can be good enough in a diffusive medium. Now, with direct synthesis, this antenna takes the central stage, as it becomes a wavefront synthesizer. A single antenna can thus potentially generate any wavefront just by acting on the signals applied to it. This procedure can be enacted in real time, as long as Green's functions are measured and stored. Sequences of wavefronts can therefore be generated, e.g., in order to image systems under test, as discussed in Section 2.4.6.

The price to pay is now the need to know Green's functions over this same surface. This is a delicate task, since in the case of a diffusive medium they are unpredictable. On the one hand, numerical simulations are an unlikely solution, since they would be too sensitive to tolerances in boundary conditions; even if this was not the case, they would require a precise description of the medium, the materials involved, and their geometry. Moreover, the reverberant nature of diffusive media implies long time evolutions, and thus necessarily long simulation times. On the other hand, measuring Green's functions in a diffusive medium is not necessarily a simpler task: the medium sensitivity to any boundary modifications (e.g., antennas moving, metallic parts introduced, etc.) means that electromagnetic probes would need to be very weakly perturbative in order not to alter the medium. On top of this problem, field probes would have to be moved along Σ – certainly not by hand, since potentially thousands of positions would be needed. A solution to this problem is described in Section 2.4.4, based on an automatic experimental characterization of a diffusive medium.

2.4.4 Implementing GTR

In order to translate the idea of virtual sources from theory to practice, a number of issues need to be addressed. The most important is certainly to decide what kind of diffusive media should be used. At the beginning of this chapter we stressed how common are diffusive media; why then bother to pick one? Given a set of media sharing the same statistical properties (Section 2.2.2), a sensible criterion would be to compare their energy efficiencies. As a matter of fact, if GTR were used for the conversion of signals into wavefronts and these wavefronts were meant for the radiative testing of a device or system, it is natural to prefer a medium that would generate the strongest possible fields for the same peak input power (Section 2.3.5). This preference takes its full sense when thinking of tests on non-linear devices, or in general in the case of devices with a threshold behavior: if an impinging field strong enough makes a device switch from an acceptable behavior to one where its basic workings is no longer ensured, it is important to have the ability to reach and test this region in the response of the device under test.

The most energy-efficient diffusive medium is the reverberation chamber, thanks to its reflective boundaries, defining a closed space where waves propagate through multiple scattering events. Low dissipation over these boundaries and an even lower rate of energy leakage make reverberation chambers the natural choice to implement GTR.

However, the advantages of reverberation chambers come with drawbacks: featuring long time constants, due to the long number of times waves propagate through them before losing a substantial fraction of their energy, they are inevitably very sensitive to any modification along their propagation paths. Displaced boundaries, unless very weakly scattering, can have a dramatic impact on the transfer functions of a reverberation chamber. Modifications like these are an issue only in one case: if a transfer function is measured before (nominal case) and after the chamber is modified. If the nominal response were to be used for the prediction of the actual chamber response, its inaccuracy could make it useless. This is the second issue that needs to be solved: how can Green's functions be measured in a medium that is so sensitive to modifications?

The solution comes in several steps. First, rather than applying elementary sources over Σ and measuring received voltages at the TRM antenna, the reciprocity of the medium can be assumed and exploited by turning the approach upside-down: the TRM antenna is excited and the field it generates over Σ is now sampled by probes. The main difference in these two opposed scenarios is that the TRM antenna is already present, and thus does not alter the medium, as opposed to the case where sources would be placed over Σ. Moreover, very weakly scattering probes are available off the shelf, whereas elementary sources are more of a fiction. It goes without saying that these probes need to be phase sensitive. Manufacturers producing this kind of probe are Enprobe, Kapteos, and Seikoh-Giken; this list is not exhaustive.

If electric- and magnetic-field probes are available, then (2.55) can be directly implemented. However, in practice magnetic-field probes are less sensitive and could have issues trying to properly measure the magnetic field generated by low-power sources, e.g., by a vector network analyzer. In such scenarios it is therefore necessary to modify (2.56) in order to have only Green's electric–electric function appearing. Demonstrating the feasibility of this idea is not trivial; the interested reader is invited to refer to [46] for a detailed discussion.

In the context of this chapter, we will stick to an intuitive explanation. Suppose another surface is to be added to Σ, concentric to it but with a radius larger by an amount ΔR. The two surfaces are now referred to as $\Sigma_l, l \in [1, 2]$. The electromagnetic field radiated by the virtual source is sampled over these two auxiliary surfaces, yielding two sets of equivalent electric current distributions. If two points over the auxiliary surfaces are chosen to be along the same radial direction from the surfaces center, the two current elements thus selected can be regarded as forming a two-element end-fire array. In this respect, weights can be applied to them in order to maximize the ratio between the outward and inward radiation. This idea can be formalized as done in [46], and shown to lead to

$$N_{\text{eq}}(\mathbf{r}) = -\zeta_o^{-1} \sum_{l=1}^{2} A_l N_{e,l}(\mathbf{r}) \qquad (2.57)$$

as a replacement to (2.56), with $N_{e,l}(r)$ the electric field generated by the TRM antenna, as sampled over the surface Σ_l, and

$$A_1 = -\frac{\exp(-2jk\Delta R)}{1 - \exp(-2jk\Delta R)}$$

$$A_2 = -A_1 \exp(2jk\Delta R)$$

(2.58)

are the weights applied to each layer of current.

Equipped with this updated formulation, the last remaining obstacle is the need to scan each auxiliary surface automatically. A solution has been presented in [47], based on a dielectric field scanner. The hardest problem in defining an automatic solution is to avoid modifying the propagation medium as a probe is moved over the surfaces. Clearly, it is not so much the probe that is to blame, but rather the inevitable mechanical support meant to move it. Standard solutions based on metallic structures covered by absorbers are clearly not fit for the task, since they would maximally modify the behavior of a reverberation chamber.

Figure 2.16 shows two pictures of the robot developed in CentraleSupelec for this purpose. It is made of strands of fiberglass held together by a polymeric matrix. The two semicircular structures are hollow, 4 cm wide, with a thickness of 1 mm. They have sufficient rigidity to ensure a precision of the end position of the probe within 1 mm. Two arms are used, one fixed, along φ, the other mobile, along ϑ.

(a)   (b)

Figure 2.16 The dielectric robot (a) with two dielectric arms and its metallic platform hosting the motors and (b) a close-up of the arms crossing, showing the belts and probe holder.

The reason for choosing two structures rather than just one is to be found in the limited rigidity, or (better) Young's elastic modulus, of plastic materials, though it is remarkably high with respect to similar materials. The arch along the φ direction dramatically helps in reducing the deflection in the mobile arch, introducing a fixed support halfway along the mobile arch. Moreover, the fixed arch also offers a simple solution to the problem of putting in motion the mobile arch; directly applying a torque to the ends of the mobile arch is not an option, as it would require over-large ends to stand the intense strain forces. It is rather simpler and more effective to have the mobile arch pushed and pulled by its middle point, where the two arms meet. A polymeric belt is used to this effect, as visible in Figure 2.16b. A similar solution is used to move a probe holder guided along the mobile arch.

An experimental analysis of the perturbation introduced by this structure has shown that for an empty chamber the field scanner is practically invisible up to 1 GHz and starts to have a substantial impact on transfer functions (and thus Green's functions) above 3 GHz. As discussed earlier, if the number of times waves travel across the chamber were to be reduced, the perturbation would be better controlled. In practice, this amounts to increased dissipation, e.g., by introducing absorbers within the chamber. As discussed in [47], a single absorber pan would lead to an almost invisible scanner up to 3 GHz.

2.4.5 Retrieving Free-Space Conditions: Experimental Results

The ideas introduced so far are now tested against experimental results. The main goal is to image the wavefronts generated with GTR, using the dielectric field scanner described in the previous section; theoretical predictions of wavefronts propagating in free-space conditions will serve as references in order to assess the accuracy of GTR.

With reference to Figure 2.17, all wavefronts are chosen to share the same structure

$$E_{wf}(r, v) = \hat{\varphi} G(r - r_s, v) \exp\left(-\psi^2/2\psi_s^2\right) P(v) \qquad (2.59)$$

where $G(r, v)$ is the scalar Green function for free-space propagation, r_s is the position of the virtual source, and ψ_s controls

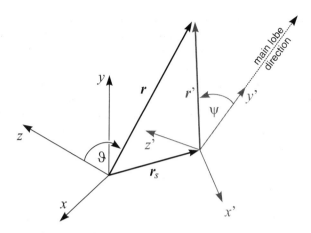

Figure 2.17 Global and local reference frames for the computation of the vector field distribution at r, radiated by a source centered over r_s.

the angular divergence of the wavefront. The time evolution of the wavefront, which is usually expected to follow a short pulse in applications of time reversal, is described by the Fourier spectrum $P(v)$, defined as

$$P(v) = \exp\left[-(v - v_c)^2/2v_s^2\right] \qquad (2.60)$$

with v_c the central frequency of the pulse and v_s its equivalent bandwidth.

In (2.59), the polarization of the wavefront is set by default along $\hat{\boldsymbol{\varphi}}$, while the wavefront has a maximum of radiation along the y axis, where ψ approaches zero. In fact, this description refers to the local reference frame marked by primed quantities in Figure 2.17. In a general way, the wavefronts associated with (2.59) will be defined by local parameters, e.g., B_e, v_c, ψ_s, and global ones, which describe how the local frame is oriented with respect to the global one. In this way, the main direction of propagation of the wavefronts can be changed, together with its co-polarization, by simply introducing a sequence of elementary rotations with respect to the global frame axis.

Equation (2.59) is well-suited to describe the field radiated by the virtual source, but it cannot be used directly to compute the space–time evolution of the refocusing wavefront, which is expected to be generated by GTR thanks to (2.55) together with (2.56) or (2.57). In order to predict the refocused field distribution, the simplest option is probably to consider the far-field radiation pattern of the virtual source, e.g., by letting $r = R\hat{k}, R \to \infty$, along a generic direction $\hat{k}$, an operation that corresponds to obtaining the plane-wave spectrum of the radiated field [23, 48]. It is then straightforward to compute the refocused field distribution, by phase-conjugating (i.e., time-reversing) the plane-wave spectrum

$$E_{GTR}(\boldsymbol{r}, t) = \int d\nu \, e^{j2\pi\nu t} \int d\hat{k} \, E_{wf}^{*}(R\hat{k}, \nu) e^{jk_o\hat{k}\cdot\boldsymbol{r}} \qquad (2.61)$$

It is worth recalling that the field scanner described in Section 2.4.4 can only measure the transfer functions between the TRM antenna and the auxiliary surfaces $\{\Sigma_l\}$. While sufficient for the synthesis of the excitation signals, these data are of limited importance in the validation of GTR. This conclusion can be understood in the light of the existence of a fluctuation background in time-reversal applications in diffusive media (see Section 2.3.2); only where wavefronts focus can they be stronger than the background. Unless a wavefront is made to focus close to the auxiliary surfaces, its intensity will practically be negligible with respect to the fluctuations. A validation of GTR must necessarily be sought around the focusing region.

Figure 2.18 shows the planar region that was manually scanned with an electric-field probe while exciting the TRM antenna, in the same way as the scanner did over the hemispherical surface. The fixed arm of the scanner is shown for reference.

Calling these transfer functions $\boldsymbol{\Pi}(\boldsymbol{r}, \nu)$, the field generated by GTR can now be computed as

$$E(\boldsymbol{r}, t) = \int d\nu \, \boldsymbol{\Pi}(\mathbf{r}, \nu) V^{*}(\nu) e^{j2\pi\nu t} \qquad (2.62)$$

where $V^{*}(\nu)$ is the phase-conjugated version of (2.55). The field distributions over the planar region are computed from frequency-domain data, since otherwise the measurements would have had to be measured for each wavefront generated. Since the planar region is manually scanned, a time-domain approach would have been too time-consuming.

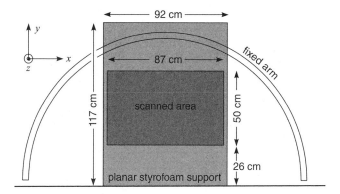

Figure 2.18 Layout of the scanned planar area as part of the xy plane of the global reference frame in Figure 2.17.

The results shown below refer to $\psi_s = 40$ degrees and $B_e = 0.5$ GHz, i.e., a strongly focusing wavefront, confined both in time and space. The theoretical and experimental distributions for a wavefront focusing at the center of the planar region are shown in Figure 2.19, for the two Cartesian field components

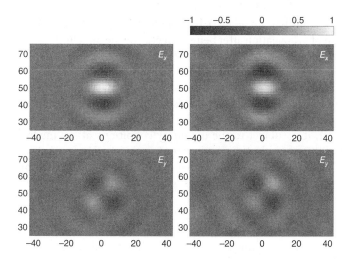

Figure 2.19 Distributions of the horizontal (E_x) and vertical (E_y) field components expected (left column) and measured (right column) at $t = 0$ ns.

lying on the plane. Only the instant $t = 0$ is shown, i.e., the time of focusing. The two distributions closely agree and each scalar component of the electric field is well reproduced, as could have been expected from Section 2.3.4.

Restraining to the main polarization component, i.e., E_x, the time evolution of the GTR-generated wavefront in Figure 2.20 can be compared to the theoretical one in Figure 2.21, from -5 ns up to 3 ns. These results show the instants during which the coherent energy starts to gather around the focus at -1.5 ns, building up the focal peak occurring around 0 ns.

The two sets of results agree very closely, confirming that a diffusive medium can indeed generate coherent wavefronts and that GTR can do it without any original source that was later time reversed.

Two instances are perhaps even more revealing. The first we focus on occurs at 3 ns. At this time the coherent wavefront we wanted to generate has finished, as can be seen in Figure 2.21. However, in Figure 2.20 it looks as though a coherent wave is still propagating. The direction of curvature of the wavefront suggests that its phase center is now well below the imaged region, and the large curvature radius implies that the phase center should be looked for below the floor. In fact, at 3 ns the coherent wavefront has already reached the floor and reversed its direction of propagation after being reflected by the metallic floor. The phase center is therefore about 50 cm below the floor. This observation is very important, since it implies that the wavefront generated with GTR is not an illusion existing only for a short period of time, but is a wavefront that is really propagating through the medium. Once coherent energy has been shaped into the wavefront, it is subject to the boundary conditions of the medium, and thus its reflections and scatterings can be observed as long as its intensity is well above that of the background fluctuations.

The second instant of interest occurs at -5 ns, when the wavefront is still outside the imaged region, as confirmed by Figure 2.21. Looking at Figure 2.20, the field is not negligible, and closely looks like the speckle distributions discussed in Section 2.3.2. This same kind of distribution, in particular its average intensity, can be observed in the background through all the time-domain results. As already pointed out in Section 2.3.2,

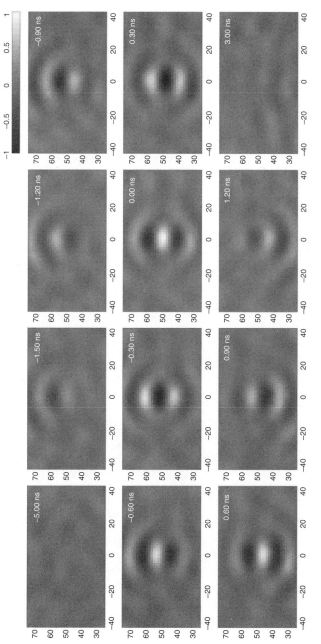

Figure 2.20 Experimental distributions of the horizontal field component E_x expected over the region shown in Figure 2.18, for $\psi_s = 40$ degrees, $B_e = 500$ MHz.

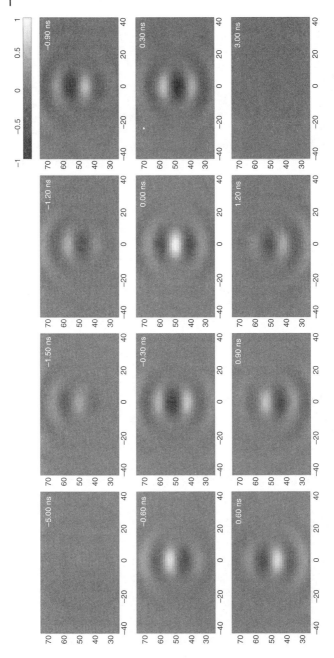

Figure 2.21 Theoretical distributions of the horizontal field component E_x expected over the region shown in Figure 2.18, for $\psi_s = 40$ degrees, $B_e = 500$ MHz.

these fluctuations are intrinsically caused by the properties of the diffusive media, and therefore cannot be removed.

The problem should rather be looked at the other way around, i.e., by looking for ways of having the coherent part of the wavefront standing out strongly enough to make background fluctuations negligible. It was recalled in Section 2.3.3 that the ratio between coherent and incoherent energies is fixed by the medium statistics. Hence, for a given background intensity, a coherent wavefront can stand out only when its energy is found within a region as compact as possible. In this case, even a small coherent energy can produce high local peaks of energy density, i.e., a peak in the electromagnetic field.

Only two options are thus available, either to focus energy in space or in time. In the first case, the plane-wave spectrum of the virtual source needs to cover the broadest spectrum of directions in order to produce a small focal spot. As is well-known from optical diffraction theory [22], this approach has its own limits, in the sense that the focal region cannot be made smaller than about half a wavelength; here the central frequency should be used for computing the wavelength. Moreover, increasing the divergence of the field radiated by the virtual source is not always possible, since it also impacts on the way a wavefront excites an equipment under test, being richer or poorer in directions of arrival.

The second option available is to increase the bandwidth of the wavefront in order to produce pulsed excitations. This approach has a limited impact on the shape of the focal spot, but again it is not always viable, as it affects the response of an equipment under test. Examples of the way these two approaches affect the accuracy (and contrast) of GTR-generated wavefronts are shown in Figure 2.22.

The results shown so far have been obtained with a single focal position and direction of arrival of the wavefront. Test facilities involving radiated fields are based on the idea that independently from the position of the source and its orientation, the wavefronts are all identical apart for rotations and translations. The invariance of the wavefronts can also be expected for GTR, since diffusive media are characterized by spatial stationarity (translation invariance) and isotropy (rotational symmetry) of their statistical properties, as argued in Section 2.1. Experimental

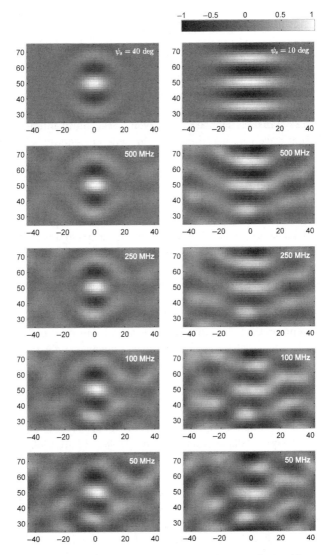

Figure 2.22 Comparison between the field distributions obtained for two Gaussian beams, with $B_e = 500$ MHz, for $\psi_s = 40$ degrees (left column) and $\psi_s = 10$ degrees (right column), as the equivalent bandwidth B_e changes. Notice how the focal region is hardly affected for $\psi_s = 40$ degrees, even for a relatively narrow bandwidth, as opposed to the case for $\psi_s = 10$ degrees.

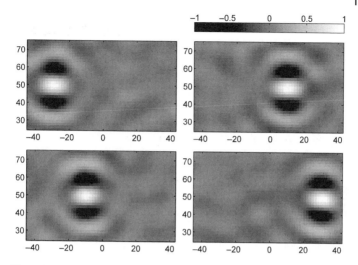

Figure 2.23 Measured field distributions of E_x, for $\psi_s = 40$ degrees, $B_e = 500$ MHz, and $t = 0$ ns, as the position of their focus is displaced horizontally by 20 cm steps.

validation of these predictions can be obtained very easily with GTR, as one just needs to change the description of the virtual source. First, spatial translations are considered. Figure 2.23 shows the field distribution generated at 0 ns by GTR when the same virtual source is shifted horizontally. The reproducibility is excellent, with no visible differences apart from the background fluctuations.

In a similar way, rotational invariance was tested, by acting on the orientation of the local reference frame in Figure 2.17. Figure 2.24 shows the results of the rotations; since with rotations the dominant role of a single Cartesian field component is lost, results are now shown as vector fields, together with a color-coded representation of the field norm. Again, the focal spot is very well preserved through the rotations.

2.4.6 An Original Application: Imaging Apertures in a Metallic Shield

Wavefronts generated with GTR can be used as excitations for any test involving electromagnetic radiations. The advantages of

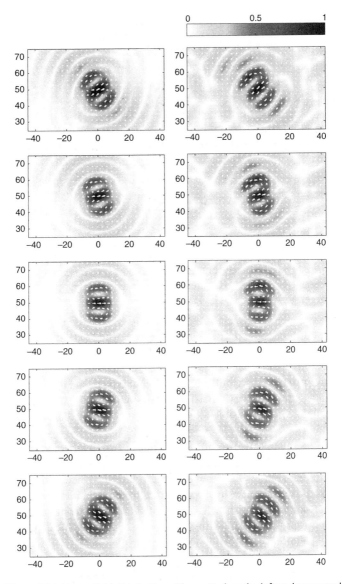

Figure 2.24 Vector field distributions (theoretical on the left and measured on the right) for $\psi_s = 40$ degrees, $B_e = 500$ MHz, at 0 ns, for wavefronts impinging along a direction of arrival $\hat{r}_{inc} \in xy$, rotated by 20 degree steps, from -40 to 40 degrees with respect to the y axis.

using GTR in diffusive media have already been stated, but two of them have a special importance, in our opinion: the ability to generate focusing wavefronts and to do it on the go. Focusing wavefronts are important since they correspond with spatial resolution, i.e., applying electromagnetic stress only to a small region of space, as opposed to diverging wavefronts.

One example of an application where these two properties are useful is the search for apertures in a metallic shield. Wavefronts generated by standard facilities, such as in anechoic chambers, have no resolution power, since they are supposed to locally mimic plane waves. As a result, when they impinge on an equipment under test, they excite a large portion of its surface at the same time: if one or more apertures are excited at the same time, it is hardly possible to distinguish their respective contributions.

GTR hands a nice solution to this problem. Focusing wavefronts can be defined and generated as a testing tool to apply stress only to a portion of the equipment surface. If an aperture exists at the position where the wavefront impacts, energy is transmitted inside the shield; it can then be measured by means either of a probe inserted for this purpose or by monitoring voltages or currents along the circuits hosted inside the shield. If the wavefront is then translated to scan the entire shield surface, step by step, an image can be produced, translating how much energy is coupled through a given region of the shield.

This idea was tested on the metallic shield in Figure 2.25. Several slots were cut through its top surface, but only one at the

Figure 2.25 Slotted metallic box imaged by scanning its top surface with focusing wavefronts.

time was left open, with the others covered with copper foil tape, in order to study the different responses of the shield when a slot has a different position. In the rest of this section only the configuration shown in Figure 2.25 is discussed.

The shield was tested by having wavefronts focusing on specific positions over the top surface of the shield, with their main lobe oriented orthogonally to it. The orientation of propagation of the wavefronts was kept unchanged through all the tests, even though it could easily be changed. Two polarizations were used in order to have the focal spot of the impinging wavefront polarized either along x or z. The central frequency was chosen to be 2 GHz, as that ensures a spatial resolution below 10 cm, according to the diffraction limit. Several bandwidths were considered, in order to show how this parameter impacts the images.

Focal spots are meant to scan the plane tangent to the shield top surface by changing the position of the associated virtual source. In order to ensure that the wavefront testing the shield at each position is seen as descending down on to the shield, it is fundamental that the wavefront corresponds to a plane-wave spectrum different from zero only for these downward directions.

The effects that these wavefronts have on the shield were measured by introducing an electric-field probe through one of the lateral surfaces of the shield, as can be seen in Figure 2.25. The peak intensity of the signals received by the probe were recorded for each position scanned, thus forming an image of the coupling through the shield surface.

Figure 2.26 shows the results of this operation. The coupling strength, measured by the peak of the signals received by the internal probe, is color-coded in order to ease the interpretation of the results. Starting with the first column (1 GHz equivalent bandwidth), the results show a coupling around the position of the only open slot, by far stronger when submitting the shield to x-polarized fields rather than z-polarized fields. This is consistent with the polarization filtering imposed by a thin slot, acting as an undercut waveguide for z-polarized fields. The region of strong coupling appears smeared outside the slot perimeter, as is inevitable due to the finite dimensions of the focal spots used during the scans: hence, even when hitting the shield about a

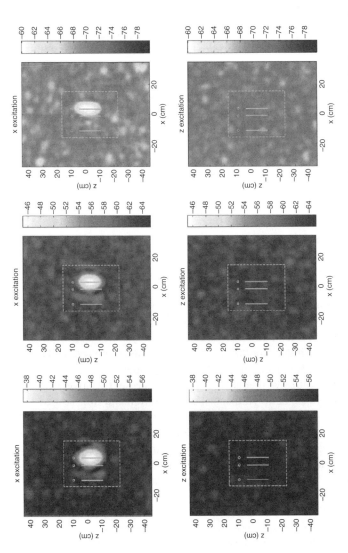

Figure 2.26 Transmission-based images of the slotted box in Figure 2.25. Results for two polarizations of the impinging wavefronts (top and bottom row), for B_e equal to 1 GHz, 312 MHz and 62 MHz, and central frequency 2 GHz. The contour of the box and the slots is superimposed on the results.

quarter of a wavelength away from the slot, a substantial portion of energy is still applied over the slot. The random distributions appearing in the background when the coupling is at its lowest are due again to background fluctuations generated by time-reversed signals within diffusive media, which is always present and thus couples in any case to the interior of the shield.

The other two columns of Figure 2.26 show how these results are modified when a narrower bandwidth is used for the temporal evolution of the test wavefronts. Substantially, the resolution is not affected. The main effect is the reduction of contrast between the slot image and the background. This reduction may become an issue if it passes below 6 dB, as at that point random fluctuations could be misinterpreted as apertures in the shield.

2.5 Final Considerations

This chapter has discussed the main reasons why diffusive media are usually considered as adverse to signal transmissions and wave propagation. It is all the more fitting that when excited by time-reversed signals these same media appear to approach the behavior of free-space environments.

This seemingly paradoxical observation has been explained by means of a self-averaging mechanism, proper to time reversal applied to diffusive media, as long as they present relatively narrow coherence bandwidths, i.e., reverberation and time spread.

Of the two groups of applications discussed, generalized time reversal is perhaps the most striking, since it seems to possess all the advantages of reverberation chambers (e.g., energy efficiency) and of anechoic ones (free-space-like propagation). In fact, what is actually surprising is the access it gives to a whole group of new properties, such as the possibility of generating arbitrary wavefronts propagating along any direction just by changing a single excitation signal on one antenna.

The application of these properties to shield imaging is a good illustration of the reasons why time-reversal excitation of diffusive media is a promising field of research, since it would have been infeasible in either anechoic or reverberating chambers.

References

1 M. Schroeder, "Statistical parameters of the frequency response curves of large rooms," *Journal of Audio Engineering Society*, vol. 35, no. 5, pp. 299–305, 1987.

2 H. Kuttruff, *Room Acoustics*. Taylor & Francis, 2000.

3 R. Vaughan and J. B. Andersen, *Channels, Propagation and Antennas for Mobile Communications*. Institution of Electrical Engineers, 2003.

4 A. Ishimaru, *Wave Propagation and Scattering in Random Media*, vol. 12. Wiley-IEEE Press, 1999.

5 J. W. Goodman, *Statistical Optics*. John Wiley & Sons, 2015.

6 P. Clemmow, *The Plane Wave Spectrum Representation of Electromagnetic Fields*. Oxford and New York: Pergamon Press, 1966.

7 T. Lehman, "A statistical theory of electromagnetic fields in complex cavities," *Interaction Notes, Note 494*, 1993.

8 D. Hill, "Electromagnetic Theory of Reverberation Chambers," National Institute of Standards and Technology, Technical Report, 1998.

9 R. Waterhouse, "Statistical properties of reverberant sound fields," *Journal of the Acoustical Society of America*, vol. 35, p. 1894, 1963.

10 R. V. Waterhouse, "Statistical properties of reverberant sound fields," *The Journal of the Acoustical Society of America*, vol. 43, p. 1436, 1968.

11 J. Davy, "The relative variance of the transmission function of a reverberation room," *Journal of Sound and Vibration*, vol. 77, no. 4, pp. 455–479, 1981.

12 R. Langley and A. Brown, "The ensemble statistics of the energy of a random system subjected to harmonic excitation," *Journal of Sound and Vibration*, vol. 275, no. 3–5, pp. 823–846, 2004.

13 K. S. Stowe, *Introduction to Statistical Mechanics and Thermodynamics*. New York: John Wiley & Sons, Inc., 1984.

14 A. Cozza, "The role of losses in the definition of the overmoded condition for reverberation chambers and their statistics," *IEEE Transactions on Electromagnetic Compatibility*, vol. 53, no. 2, pp. 296–307, 2010.

15 J. Philibert, "One and a half century of diffusion: Fick, Einstein, before and beyond," *Diffusion Fundamentals*, vol. 4, no. 6, pp. 1–19, 2006.

16 D. Hill and J. Ladbury, "Spatial-correlation functions of fields and energy density in a reverberation chamber," *IEEE Transactions on Electromagnetic Compatibility*, vol. 44, no. 1, pp. 95–101, 2002.

17 A. Derode, P. Roux, and M. Fink, "Robust acoustic time reversal with high-order multiple scattering," *Physical Review Letters*, vol. 75, no. 23, pp. 4206–4209, 1995.

18 A. Derode, A. Tourin, and M. Fink, "Random multiple scattering of ultrasound. II. Is time reversal a self-averaging process?" *Physical Review E*, vol. 64, no. 3, p. 36606, 2001.

19 A. Derode, A. Tourin, J. de Rosny, M. Tanter, S. Yon, and M. Fink, "Taking advantage of multiple scattering to communicate with time-reversal antennas," *Physical Review Letters*, vol. 90, no. 1, p. 14301, 2003.

20 C. Draeger, J. Aime, and M. Fink, "One-channel time-reversal in chaotic cavities: experimental results," *The Journal of the Acoustical Society of America*, vol. 105, p. 618, 1999.

21 M. Fink and C. Prada, "Acoustic time-reversal mirrors," *Inverse Problems*, vol. 17, p. R1, 2001.

22 M. Born and E. Wolf, *Principles of Optics*. New York: Pergamon Press, 1980.

23 R. Collins and F. Zucker, *Antenna Theory*. New York: McGraw-Hill, 1969.

24 A. Cozza, "Statistics of the performance of time reversal in a lossy reverberating medium," *Physical Review E*, vol. 80, no. 5, p. 056604, 2009.

25 A. Cozza and F. Monsef, "Multiple-source time-reversal transmissions in random media," *IEEE Transactions on Antennas and Propagation*, vol. 62, no. 8, pp. 4269–4281, August 2014.

26 H. Vallon, A. Cozza, F. Monsef, and A. Chauchat, "Time-reversed excitation of reverberation chambers: improving efficiency and reliability in the generation of radiated stress," *IEEE Transactions on Electromagnetic Compatibility*, vol. 58, no. 2, pp. 364–370, 2016.

27 K. Sarabandi, I. Koh, and M. Casciato, "Demonstration of time reversal methods in a multi-path environment," in *IEEE*

International Symposium of the Antennas and Propagation Society, vol. 4, pp. 4436–4439, IEEE, 2004.

28 P. Kyritsi and G. Papanicolaou, "One-bit time reversal for WLAN applications," in *PIMRC 2005,* vol. 1, pp. 532–536, IEEE, 2005.

29 R. Qiu, C. Zhou, N. Guo, and J. Zhang, "Time reversal with MISO for ultrawideband communications: experimental results," *Antennas and Wireless Propagation Letters,* vol. 5, no. 1, pp. 269–273, *IEEE,* December 2006.

30 H. T. Nguyen, J. B. Andersen, G. F. Pedersen, P. Kyritsi, and P. C. F. Eggers, "Time reversal in wireless communications: a measurement-based investigation," *IEEE Transactions on Wireless Communications,* vol. 5, no. 8, pp. 2242–2252, 2006.

31 P. Pajusco and P. Pagani, "On the use of uniform circular arrays for characterizing UWB time reversal," *IEEE Transactions on Antennas and Propagation,* vol. 57, no. 1, pp. 102–109, 2009.

32 I. Vellekoop, A. Lagendijk, and A. Mosk, "Exploiting disorder for perfect focusing," *Nature Photonics,* vol. 4, no. 5, pp. 320–322, 2010.

33 A. Cozza and H. Moussa, "Enforcing deterministic polarisation in a reverberating environment," *Electronics Letters,* vol. 45, no. 25, pp. 1299–1301, 2009.

34 D. Hill, "Electronic mode stirring for reverberation chambers," *IEEE Transactions on Electromagnetic Compatibility,* vol. 36, no. 4, pp. 294–299, November 1994.

35 H.-J. Stöckmann, *Quantum Chaos: An Introduction.* Cambridge University Press, 2006.

36 A. Cozza and H. Moussa, "Polarization selectivity for pulsed fields in a reverberation chamber," in *2010 Asia-Pacific Symposium on Electromagnetic Compatibility (APEMC),* pp. 574–577, IEEE, 2010.

37 A. Cozza, "Increasing peak-field generation efficiency of reverberation chamber," *Electronics Letters,* vol. 46, no. 1, pp. 38–39, 2010.

38 C. Draeger and M. Fink, "One-channel time-reversal in chaotic cavities: theoretical limits," *Journal of the Acoustical Society of America,* vol. 105, p. 611, 1999.

39 H. Moussa, A. Cozza, and M. Cauterman, "Directive wavefronts inside a time reversal electromagnetic chamber," in

    IEEE International Symposium on Electromagnetic Compatibility, 2009 (EMC 2009), pp. 159–164, August 2009.

40 C. Oestges, A. Kim, G. Papanicolaou, and A. Paulraj, "Characterization of space–time focusing in time-reversed random fields," *IEEE Transactions on Antennas and Propagation*, vol. 53, no. 1, pp. 283–293, 2005.

41 R. Harrington, *Time-Harmonic Electromagnetic Fields*. New York: McGraw-Hill, 1961.

42 A. Cozza, "Emulating an anechoic environment in a wave-diffusive medium through an extended time-reversal approach," *IEEE Transactions on Antennas and Propagation*, vol. 60, no. 8, pp. 3838–3852, 2012.

43 J. de Rosny and M. Fink, "Overcoming the diffraction limit in wave physics using a time-reversal mirror and a novel acoustic sink," *Physical Review Letters*, vol. 89, no. 12, p. 124301, 2002.

44 C. Prada and M. Fink, "Eigenmodes of the time reversal operator: a solution to selective focusing in multiple-target media," *Wave Motion*, vol. 20, no. 2, pp. 151–163, 1994.

45 C. Prada, J.-L. Thomas, and M. Fink, "The iterative time reversal process: analysis of the convergence," *The Journal of the Acoustical Society of America*, vol. 97, no. 1, pp. 62–71, 1995.

46 A. Cozza and F. Monsef, "Layered electric-current approximations of cylindrical sources," *Wave Motion*, vol. 64, pp. 34–51, 2016.

47 A. Cozza, F. Masciovecchio, C. Dorgan, M. Serhir, F. Monsef, and D. Lecointe, "A dielectric low-perturbation field scanner for sensitive environments," *IEEE Transactions on Antennas and Propagation*, 2016, under review.

48 T. Hansen and A. D. Yaghjian, *Plane-Wave Theory of Time-Domain Fields*. IEEE Press, 1999.

3

From Electromagnetic Time-Reversal Theoretical Accuracy to Practical Robustness for EMC Applications

P. Bonnet, S. Lalléchère, and F. Paladian

Université Clermont Auvergne, UBP, Institut Pascal, Clermont-Ferrand, France

3.1 On the Interest of Time Reversal in the EMC Context

This chapter is devoted to the application of the time reversal (TR) concept to the development of the methodologies in the electromagnetic compatibility (EMC) context.

For more than twenty years, EMC has constituted a distinct discipline. The nature and the characteristics of the electromagnetic sources are subject to constant change and the risks of generating some disturbances of electronic devices must without fail be considered by the designers. Indeed, the evolution of technologies is associated with a miniaturization of the internal equipment where functioning components performing at very low energy levels must operate in spite of disturbance signals. To these volume constraints, the use of composite materials allowing a weight reduction does not yet guarantee an electromagnetic mitigation for an efficient shield of external perturbations. The main objectives of the 2014/30/UE EMC European

Electromagnetic Time Reversal: Application to Electromagnetic Compatibility and Power Systems, First Edition. Edited by Farhad Rachidi, Marcos Rubinstein and Mario Paolone.
© 2017 John Wiley & Sons, Ltd. Published 2017 by John Wiley & Sons, Ltd.
Companion Website: www.wiley.com/go/rachidi55

Directive [1] are therefore to guarantee the operation of electronic devices immersed in a disturbing electromagnetic environment with a reduction of the parasitic levels generated by these same systems. These aims nowadays become a real challenge, which must be taken up despite the complexity of the current technologies. Indeed, the risk associated with an electronic failure can lead not only to damages on the system but also edge problems of equipment security and consequently human safety. For example, we can mention the field of air and land transportation but also energy supply lines and distribution installations that are being addressed by a great number of research programs.

These EMC aspects must be considered right from the design phase, implying in the case of complex systems control of the three main factors of the problem:

– The identification of the electromagnetic perturbing sources.
– The evaluation of the physical phenomena occurring from the sources up to the sensitive system: electromagnetic analysis for the electromagnetic coupling study.
– The sensitive components of the studied system: critical analysis of the functions that guarantee its operation.

The objectives of the research works in the EMC area are to develop theoretical models or measurement protocols, as well as new approaches to overcome scientific and technical obstacles that are often due to the complexity of the electronic systems. Some of the concepts recently explored include the time-reversal (TR) technique leading to innovative results in different areas.

The electromagnetic coupling depends on the physical phenomena origins. Conduction is referred to the propagation of an electromagnetic disturbance along a physical support and emission corresponds to the case of interferences produced by the electric field, the magnetic field, or the electromagnetic field characterizing the source. For the coupling evaluation, the cable networks play an important role since they are in a significant number of transportation systems (automotive vehicles, aircrafts, etc.). They are submitted to electromagnetic effects produced by direct injection of perturbing currents and voltages or by coupling with an external electromagnetic field. In most cases, the networks are composed of multiconductor cables and hence cross-talk must also be taken into account. In addition, as induced currents and voltages correspond to a wide frequency range, the electric lengths of the structures lead to consideration of propagation phenomena. Hence, the signals transported by the cables are submitted to reflexion phenomena at the conductor ends and the propagation media can be considered as a multipath environment. These high-frequency phenomena may affect the integrity of the transmitted signals and power but help to allow the application of the TR concept at different levels:

- For the identification of defects on cables, an aspect detailed in Section 3.2.
- For a focalization of the transmitted power at the equipment location while reducing the emission effects produced by the cables, for example in the case of current power line (CPL) systems [2].

To determine the currents and voltages induced by an external electromagnetic wave or by injection of a perturbing signal, the multiconductor transmission line (MTL) method, under the quasi-TEM assumption, is well adapted to the analysis of complex cable networks interconnected by devices modeled by lumped circuits [3]. Indeed, a global approach allows all the conductors of the considered configuration to be taken into account and hence all the electromagnetic couplings. Moreover, the MTL method, by allowing modeling of shields characterized by their transfer impedances and admittances, has the advantage to evaluate the impact of protective devices on the electronic end circuits. Associated with reasonable simulation times, the satisfactory comparisons between theoretical and experimental results obtained for complex cable networks have really demonstrated the great interest of this approach for the EMC study of uniform and non-uniform transmission lines.

For the electromagnetic analysis of a system, the choice of modeling the entire configuration by including all geometrical details may be a good way, implying the use of *full-wave* methods. These methods are based on resolution of the Maxwell system by different numerical techniques [4], such as integral methods and the moment method (MoM), the boundary element method (BEM), the PEEC method (where the partial element equivalent circuit does not taking emission phenomena into

Figure 3.1 MSRC of Institut Pascal (8.40 m × 6.73 m × 3.50 m).

account), and differential methods: the finite element method (FEM), the finite integration technique (FIT), and the finite difference (FD). If a prototype of the considered equipment or system exists, specific EMC experimental devices allowing plane-wave generation may be complementary to a theoretical approach. To perform radiated immunity or emission tests, we can mention anechoic chambers, i.e., Faraday cages where walls are equipped to absorb materials, TEM and GTEM cells, striplines, etc. [5]. An alternative way is the mode stirred reverberation chamber (MSRC, Figure 3.1), operating on the physical principle of resonant metallic cavities [6]. Corresponding to a Faraday cage equipped with a stirrer, a large metallic structure in rotation in the case of a mechanical stirring, one can predefine a so-called *working volume* where the electromagnetic field is statistically, over a full rotation of the stirrer, homogeneous and isotropic. The electromagnetic field distribution is variable from one stirrer position to another, the role of the stirrer being to artificially modify the boundary conditions of the cavity. This optimal operation is estimated by means of various criteria that allow determination of a minimal frequency value in relation to the number of modes excited inside the structure. Due to reflexion and scattering phenomena on the walls and the stirrer, the system under test is submitted to a multipath environment, hence enabling conditions for application of the TR concept, as illustrated in Section 3.3.

Consequently, some of the physical phenomena and system properties being considered in EMC problems benefit from applications of the TR concept that have represented real technological advances for identification of defects in cable networks or for characterizing electronic systems in the MSRC. Such aspects have contributed to improvements in risk control of devices described in the two following sections. As explained in Section 3.2, cables represent a propagation medium well adapted to the principle of the EMTR concept, which consequently can be applied to wire diagnosis in the case of soft defects, which are particularly difficult to detect using conventional techniques. For electronic systems, EMC immunity analysis in MSRC and the application of TR methods associated with statistical approaches constitute an innovative methodology, as presented, respectively, in Sections 3.3 and 3.4. Finally, concluding remarks are proposed in Section 3.5.

3.2 TR in Transmission Line (TL) Networks

3.2.1 Transmission Line and Time Reversal

Under the quasi-TEM mode assumption [3], transmission line modeling is generally a one-dimensional approximation of the physical model, which represents the voltage and current waves as a function of the transmission line axis. The time-dependent classical Telegrapher's equations describing the wave propagation along a two-conductor (or a single wire above a perfectly conducting ground) line are given by

$$\frac{\partial v(x,t)}{\partial x} + L\frac{\partial i(x,t)}{\partial t} + Ri(x,t) = 0 \tag{3.1}$$

$$\frac{\partial i(x,t)}{\partial x} + C\frac{\partial v(x,t)}{\partial t} + Gv(x,t) = 0 \tag{3.2}$$

In which $v(x,t)$ is the line voltage at position x along the transmission line axis at time t, $i(x,t)$ is the line current and L, C, R, G are the per-unit-length (p.u.l.) parameters, respectively, for: the series resistance (Ω/m), the series inductance (H/m), the shunt capacitance (F/m), and the shunt conductance (S/m). Note that multiconductor or bundle cables are characterized by p.u.l. parameter matrices. Telegrapher's equations can be solved by various numerical methods (for instance, a finite difference time-domain method, FDTD [7].

For lossless (non-dissipative) transmission lines (i.e., $R = 0 = G$) we notice that the combination of partial differential equations (3.1) and (3.2) leads to time-invariant wave equations for both voltage (3.3) and current (3.4):

$$\frac{\partial^2 v(x,t)}{\partial t^2} = \frac{1}{LC}\frac{\partial^2 v(x,t)}{\partial x^2} \tag{3.3}$$

$$\frac{\partial^2 i(x,t)}{\partial t^2} = \frac{1}{LC}\frac{\partial^2 i(x,t)}{\partial x^2} \tag{3.4}$$

It is clear that the voltage and the current and their time-symmetric values (i.e., $v(x,t)$ and $v(x,-t)$ or $i(x,t)$ and $i(x,-t)$) are both solutions of the same propagation equation. To make Telegrapher's equations (3.1) and (3.2) also invariant under time reversal ($t \rightarrow -t$), the current should change sign as well $i(x,t) \rightarrow -i(x,-t)$:

$$\frac{\partial v(x,-t)}{\partial x} + L\frac{\partial(-i(x,-t))}{\partial(-t)} = 0 \tag{3.5}$$

$$\frac{\partial(-i(x,-t))}{\partial x} + C\frac{\partial v(x,-t)}{\partial(-t)} = 0 \tag{3.6}$$

Indeed, the mathematical time-reversal operation implies changing the sign of the charge velocity and consequently the sign of the current.

Hence the concept of TR can be applied to lossless and invariant transmission lines or a transmission lines network (TLN). According to the wave equations or Telegrapher's equations, the reverse of the voltage (or the current) in the time domain (or phase conjugation in the frequency domain) would precisely retrace the path of the original wave back to the source generating the signal. From a theoretical point of view, the TR operator should be applied to the voltage at any point of the TLN. Unfortunately this cannot be done in practice. The same difficulty appears to apply to the TR principle for Maxwell's equations in three dimensions. To overcome this limitation one can refer to the use of the Huygens principle: the electromagnetic field at any point of the domain can be computed from the knowledge of the fields on a closed surface surrounding the sources inside the volume defined. Then the time-reversal procedure consists in recording the electromagnetic signal during a limited duration, time-reversed. and then re-transmitted into the medium. From an experimental point of view, it remains impossible to measure and re-emit the field at any point on a two-dimensional surface. Thus, for practical reasons, the time-reversal procedure is only performed on a limited angular area (the so-called time-reversal mirror, TRM) where the field is spatially sampled (Figure 3.2a). One can imagine that this procedure leads necessarily to deterioration in the TR properties as will be observed in Section 3.3.

Alternately, in a transmission line problem, neglecting the wire's radiated electromagnetic field, a perfect experimental TRM based on all line terminations can be achieved (Figure 3.2b). Assume a transmission line network with a localized source of voltage emitting a signal that propagates throughout the network. Consider digital oscilloscopes at each end of the network that store the associated signals. If these signals are re-emitted back in the network after time reversal thanks to the same number of arbitrary waveform generators, the resulting voltages will converge on to the source that originally created the voltage. In this purely one-dimensional problem all the information propagating in a TLN can a priori be easily caught, which is a favorable configuration for potential time-reversal applications.

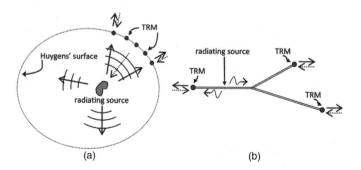

Figure 3.2 Schematic representation of the TRM principle: (a) in a three-dimensional problem; (b) in a transmission line network.

3.2.2 Defect or Fault Detection in Transmission Lines

Wiring diagnosis is an important issue in EMC since some cable's alterations (localized damage to the insulation or shielding of a wire) can dramatically modify the coupling. Open and short circuits are extreme modifications which have to be detected at an early stage when small anomalies in cable impedance appear. The former cases, known as hard defects, that produce large signal reflection are quite easy to deal with. The latter that cause small impedance changes, and thus small reflections, are often confused by the noise of measurement and are called soft defects (Figure 3.3).

Among all known diagnosis methods, the most widely used approaches for wire defect detection and location are reflectometry-based techniques. An excellent and interesting review can be found in [8]. In a general way, two main types of reflectometry are considered: time-domain reflectometry (TDR) and frequency-domain reflectometry (FDR). The nature of the test signal used, a Gaussian pulse or voltage step for TDR and a set of stepped sine waves for FDR, allows classification of the different approaches. Mainly due to their easier implementation and interpretation, TDR techniques are becoming more popular than FDR techniques.

In the standard time-domain reflectometry a high-frequency pulse signal is sent down the wire and then the reflected signal, sensitive to impedance variations along the wire, is analyzed. In Figure 3.4, a voltage Gaussian pulse is launched down in a simple matched transmission line of 10 cm with an impedance discontinuity located in the middle. This discontinuity generates a reflected signal recorded at the injection point. A so-called reflectogram is then obtained (Figure 3.5, upper). Knowing or evaluating the voltage wave propagation velocity enables the reflectogram to be expressed in meters (or centimeters in this

Figure 3.3 Example of a soft defect; frays in the insulation.

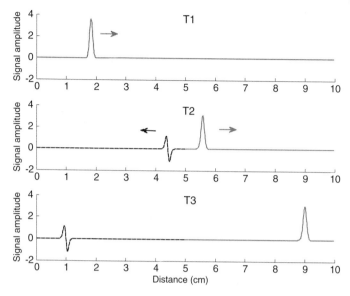

Figure 3.4 Standard TDR principle. A probe signal is sent down the cable (time T1); a part of the energy is reflected by an impedance discontinuity (time T2); and the reflected signal comes back to the injection port (time T3).

example) and consequently enables the discontinuity or defect to be located (Figure 3.5, lower).

The amount of signal reflected back is calculated by the reflection coefficient

$$\Gamma = \frac{Z_L - Z_0}{Z_L + Z_0} \tag{3.7}$$

where Z_0 is the characteristic impedance of the cable and Z_L is the impedance at the discontinuity. The reflection coefficient equals -1 for a short circuit and 1 for an open circuit, which allows quite easy identification of these hard defects. However, soft defects, like damage in the shielding of cables, frays in the insulation, and local geometrical or electrical property modifications, are by definition weakly reflective and cause small reflection coefficients ($\Gamma \ll 1$). The detection of this type of defect can become very hard in more complex topology networks

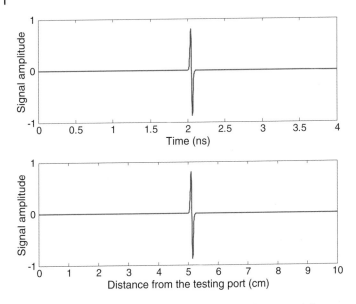

Figure 3.5 Reflectograms at the testing port: temporal representation (upper illustration) and spatial representation (lower illustration).

(Figure 3.6). TLN with junctions between different branches or mismatched load impedances create numerous reflected waves (or echoes) comparable to the defect level. In order to remove these intrinsic echoes, most TDR techniques work with the difference between the refloctograms obtained from the TLN under test and a reference reflectogram (or baseline) that is assumed to be healthy (without any defect). Although this procedure allows highlighting weak reflection related to a defect, traditional TDR techniques are not able to deal efficiently with the problem of soft defect detection. Added to that is the fact that TDR techniques are sensitive to noise and are not suitable in the case of multiple defects.

3.2.3 Time-Reversal Wire Diagnosis

The fundamental property of focusing offers by TR has been demonstrated to hold promising technological potential. In the problem of target detection in radar application, it has been

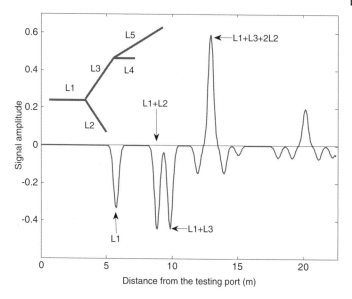

Figure 3.6 Reflectogram for a double Y-shaped network (schematic in the top left) with L1 = 5 m, L2 = 3 m, L3 = 4 m, L4 = 2 m, L5 = 5 m ($Z_0 = 50\ \Omega$) and with open circuits at line ends.

proven that TR signal processing can amplify the reflected echo from the target. Sources of radiation, or scatterers acting as secondary sources, are identified by propagating back the electromagnetic field they have radiated in the first step of the TR process. This concept can be transposed within the framework of TLN where defects (or faults) play the role of target.

Contrary to the open space propagation case, as explained previously, a perfect experimental TRM can be achieved in guided media since all the information can be captured (e.g., for lossless transmission lines and neglecting the radiation phenomena). In practice, all the terminations of a TLN are not accessible and it can be relatively expensive to use the equivalent number of arbitrary wave generators. Nevertheless, the focusing property still remains for only one element for the TRM and may take benefits from the presence of multipaths in complex networks with several junctions. This remarkable property has been demonstrated and exploited in three dimensions, as shown in Section 3.3.

The main drawback is ignorance of the focusing time since the location of the defect is unknown and, unfortunately, due to multipaths, similar peaks can appear in different locations and at different instants. If the second step of the TR process is performed numerically solving Telegrapher's equations (3.1) and (3.2) associated with the reference TLN, one can observe the voltage on each point of the network and at each time step. However, this one simulation is not enough for defect detection and localization and signal processing should be required.

Consider once again the previous example of a simple matched transmission line. If the reflectogram is time reversed and reinjected in the transmission line without any defect it would freely propagate along all of the line (Figure 3.7). Simultaneously, if we artificially and numerically back-propagate the initial incident voltage peak, these two signals will overlap at the

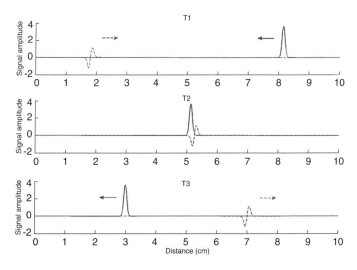

Figure 3.7 Numerical simulation of TR propagations in a simple matched transmission line. Time T1: the time-reversed reflectogram (left dashed curve) propagates down the line, while the initial incident Gaussian voltage (right plain curve) propagates up the line. These two signals overlap at time T2 corresponding to the focusing time. Artificially and for non-causality reason, these signals continue their propagation (time T3).

location where the initial voltage peak initiated the scattered signal (Figure 3.8). This spatial position corresponds to the exact location of the defect, which can be revealed, independently of time, by a convolution product between the two previous signals. Abstractly, the convolution measures the amount of overlap between two signals. It can be thought of as a mixing operation that integrates the point-wise multiplication of one dataset with another.

Based on this idea, a basic time-reversal procedure for defect detection and localization is depicted in Figure 3.8. Let Tp be a testing port located at one end of the matched transmission line ($Z_L = Z_0$). All the test signals are respectively recorded and injected on this port when experimental measurements or numerical simulations are realized.

In a first step, the same Gaussian pulse is injected separately in the lines with and without defects. In the two cases the reflected signals are recorded over time at Tp: $V_r(\mathfrak{F}, t)$ for the reference line and $V_{rD}(\mathfrak{F}, t)$ for the line with a defect. It is worth noting that the latter is generally obtained by measurement on the potentially faulty transmission line under test. Additionally, the spatial voltage distribution along the line without defect, $V_c(x, t)$, is obtained numerically and saved for each time step considered (in practice the time is sampled).

In a second step, the recorded signals are time-reversed and re-injected through the testing port Tp in the transmission line without defect. The consecutive spatial voltage distributions along the transmission line, $V_{rc}(x, t)$ associated with $V_r(\mathfrak{F}, -t)$ and $V_{rDc}(x, t)$ associated with $V_{rD}(\mathfrak{F}, -t)$, are numerically computed.

Finally, in a last step, two products similar to convolution products are performed. For each time step, the values of $V_c(x, t)$ are multiplied by the values of $V_{rDc}(x, t)$ for each sampled spatial position x. Then a sum over all the simulation durations is achieved. The same procedure is applied for $V_c(x, t)$ and $V_{rc}(x, t)$. Each operation consists in obtaining the area overlap between the two voltages. The results reach maximal values for the components of the transmission line, which has a time delay equal to the time delay corresponding to the impedance discontinuities. A subtraction operation allows the elimination of most of the

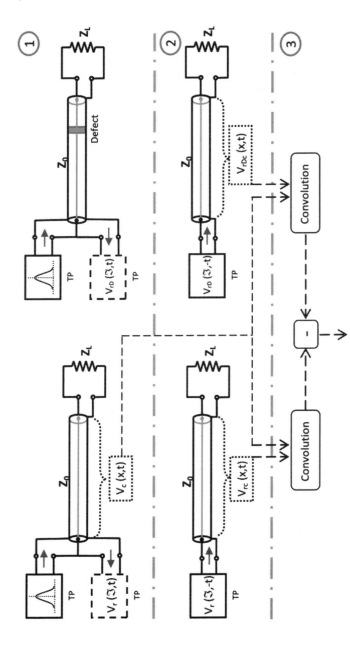

Figure 3.8 The three steps of the time-reversal procedure for defect diagnosis in a transmission line.

intrinsic echoes of the TLN that are not related to any defect (e.g., reflections at junctions or mismatched far end impedances). This *baseline approach* is an ideal procedure since in the case of multiple defects in a complex TLN, echoes initially generated by the interaction between the testing signal and the defects produce other echoes by interaction between junctions or other defects.

3.2.4   Time-Reversal Wire Diagnosis Examples

In this section, the previous TR wire diagnosis method is numerically illustrated. A simple and common way to obtain transient responses of transmission lines is to apply the wellknown FDTD technique [7] to Equations (3.1) and (3.2). The derivatives in Telegrapher's equations are then discretized both in space $(x_i = i.\Delta x)$ and time $(t_n = n.\Delta t)$ and approximated by a central finite difference scheme. Note that complete discussions concerning the FDTD method for transmission line problems can be found in Paul [3]. Although it is quite straightforward to implement a TR algorithm with an FDTD code, the fact remains that using a numerical method presents some limitations (e.g., in the case of FDTD: highest working frequency, stability criterion, numerical dispersion, etc.).

For the first example, a TLN made of one junction and three transmission lines (RG58 cable model considered as lossless) indicated by L1 (2 m), L2 (3 m), and L3 (5 m), is considered (Figure 3.9). The characteristic impedance of lines L1 and L2 is 50 Ω and 100 Ω for L3. The right end of branch L2 is opened and the branch L3 is loaded by an impedance equal to 100 Ω. The left end of line L1 is referred to as the testing port Tp where a voltage pulse of 1 V is injected. The resulting reflectogram is given in Figure 3.9 (upper illustration). As expected, the mismatched junction and the open circuit are clearly identifiable. Then, one single defect, located at 1 m from Tp on the branch L1, is tested. For that, we assume a 20% drop of the per-unit-length inductance. Due to the FDTD discretization, this soft defect is at least one spatial discretization step long. The signature of this defect is barely visible on the reflectogram (Figure 3.9, lower illustration), which fully justifies the name of soft defect (the reflection coefficient Γ is obviously very low).

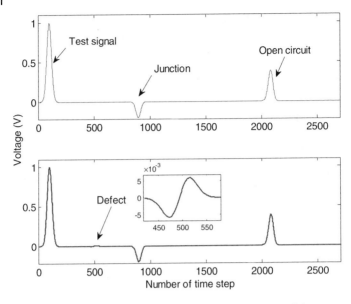

Figure 3.9 Standard reflectograms of a Y-shaped TLN without defect (upper illustration) and with defect (lower illustration).

In this example, the defect affects one FDTD cell, creating an impedance discontinuity at its left and right ends, which explains the shape of the reflected wave associated with it (zoom in Figure 3.9).

In agreement with the discussion presented in the previous section 3.2.3, the TR procedure should exhibit a maximum in correspondence to the defect location. Indeed, one peak stands out clearly and matches well with the position of the simulated defect in the TLN (Figure 3.10). It is worth noting that the higher the frequency test signals, the narrower are the peaks (Figure 3.10). For high spatial resolution, large bandwidths seem to be necessary. Therefore, accurate identification of defects would require a high-frequency pulse generator and expensive fast electronics. This should not be perceived as an issue as such, since most of the time the most important thing is to know that there is a defect and approximately its location (centimeters are sufficient enough for a posteriori visual inspection). Moreover, wider bandwidths can be an issue regarding EMC constraints or

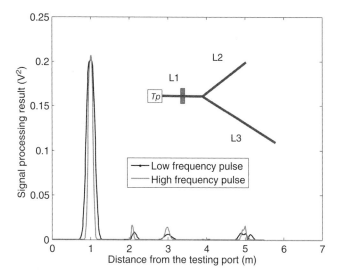

Figure 3.10 Time reversal reflectometry results. The fault is localized 1 m from the testing point *Tp*. The peak detection is obtained for two impinging pulse voltages.

in the case of a low-frequency network such as the power grid. Nevertheless, effective improved TR diagnosis methods exist to get around this problem [9].

Weaker peaks in Figure 3.10 come from the interaction of the echoes caused by the presence of a defect with the other TLN impedance discontinuities (junction, open circuit, etc.). These peaks cannot be removed with the simple baseline approach since these echoes do not exist in the healthy TLN. A more complex signal processing should be implemented.

Compared to the standard time-domain reflectometry, it has been proved that this "basic" TR wire diagnosis enhanced the amplitude of peak defect by an important factor (more than a 10 dB gain can be obtained) [10].

This improvement remains by considering multiple soft defects detection and localization. For this purpose, a second defect is added in the previous TLN. A modification of the per-unit-length parameter on line 3 at 4 m from the junction (6 m from *Tp*) is simulated. Nothing in the method prevents consideration of spread faults but, as previously mentioned, the use of

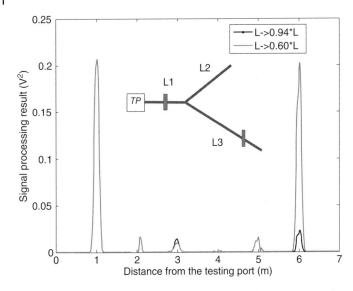

Figure 3.11 Time-reversal reflectometry results. Two faults are localized at 1 m and 6 m from the testing point *Tp*.

an FDTD scheme enforces at least one spatial cell to define the defect. This second defect is simulated by changing the value of the per-unit-length inductance by a factor of 0.60 and a factor of 0.94. The results reported in Figure 3.11 show that the peaks associated with these defects are present. Yet, in the case of the weaker defect, the amplitude of the peak is similar to those of the spurious echoes. Since these echoes are connected to the first defect and the intrinsic impedance discontinuities of the TLN, they can be removed with an ad hoc signal processing. Moreover, in the case of a noisy reflected signal, an autoconvolution step should be added after the third step in order to reduce noise. In any case, because the energy decreases since the voltage wave travels down the TLN (junctions, reflections at each defects, etc.), the detection of a defect far away from *Tp* would become increasingly difficult. Figure 3.11 illustrates this limitation: despite same peak amplitudes, the second defect severity is stronger than those of the first defect from *Tp* (modifications of 20% and 40%, respectively).

The previous examples illustrate the basic principle and the possibilities offered by TR for detecting and locating soft defects in TLN [11]. Actually, time-reversal-based techniques come to be seen as promising complementary techniques to standard TDR techniques. Since the first application of the TR concept to defect-detection in wire networks [12], various improvements have been proposed.

In essence, TR use signals determined by the TLN itself (including its topology and electrical properties). A self-adaptive definition of the signal used in the second step of the TR procedure is generated in the first step. That is the reason why some TR techniques are referred to as matched-pulse approaches [12]. In this context, similarly to acoustic [13] or three-dimensional electromagnetic propagation [14], it was demonstrated that the more complex the TLN, the more efficient the TR application is. In [15], it was also proved that frequency-dispersive defects improve the performance of time-reversal-based methods.

Also derived from acoustics [16], the expansion of the TR operator [17–19], namely the DORT method, allows electromagnetic signals expected to focus on the scatterers to be synthesized [20,21]. A differential version of the DORT method was introduced in [22] for defect-detection in wire networks. In its standard version, this DORT method does not perform well in locating multiple defects. An enhanced DORT method has been proposed that bypasses this restriction [23, 24]. It is worthy to note that DORT-based-techniques require the availability of a multiport characterization of the TLN to generate a related scattering matrix.

A TR multiple-signal classification, namely TR-MUSIC, has been recently proposed [9] for locating defect with high spatial resolution. Multiple defects (up to $N - 1$, where N stands for the number of accessible ports) can be detected with a submillimeter resolution. Moreover, this localization is obtained with relatively low-frequency continuous-wave test signals. This possibility is clearly an advantage with regard to eventual EMC constraints. Another important feature of TR-MUSIC is estimation of the defect severity. It should be noted that an original and efficient TR application to fault detection and location in power networks was also proposed [25, 26].

As well as punctual or spatially limited soft defects, global degradation in time of the electrical properties of cables can cause potential unfavorable EMC effects. In this context, but not exclusively, the evaluation of cable aging is very important and can help to anticipate a timely maintenance for degraded cables. Commonly used methods, such as time-domain reflectometry, provide non-relevant or inaccurate information in the case of overall cable aging. An alternative time-reversal-based method has been proposed in [27]. The new method, called time-reversal reflectometry, shows great ability for detecting and assessing the degree of aging. The advantages of time-reversal reflectometry over standard reflectometry is proved by experimental aging tests.

In live testing, noisy conditions are fatally present as well as losses in the TLN. In some cases, the conditions (background, electric properties, etc.) might even change between the two steps of the TR process. Section 3.4 aims to demonstrate the applicability of TR regarding real EMC experimental constraints. The next section will introduce the use of TR for EMC purposes (radiated immunity testing).

3.3 Selective EMC Radiated Immunity

3.3.1 From Open Area to Reverberating Environment for EMTR

Preliminary studies may be achieved in a free space area to assess the quality of the TR process better. In that framework, depending on the device/code used for experimental and/or numerical tests, Huygens' surface required for an ideal EMTR is approached by using a TR cavity (TRC) surrounding the area under test (i.e., the focusing area). It is well-known [28] that the number of probes greatly influences the focusing quality (maximum, signal to noise, etc.) of the time/space (data are not shown here). Despite this, since TR is very demanding regarding the number (theoretically infinity) of required probes (Huygens' surface stands for an ideal TRC), a TR mirror (TRM) is preferred with a restricted opening. In the following, time domain (FDTD) simulations are achieved in a three-dimensional (3-D) parallelepiped computational domain (CD, here $2.211 \times 1.485 \times 0.957$ m^3) with a space meshing step $\Delta = 0.033$ m in each Cartesian direction. Absorbing boundary conditions (ABCs) ensure a free-space simulation. The numerical source is a sine Gaussian pattern with central frequency $f_c = 600$ MHz with frequency bandwidth $\Delta\Omega = 350$ MHz.

Figure 3.12 gives an overview of the influence of the number of probes inside TRC in an open area. In Figure 3.12a and b more than six thousand EM field probes are given (the maximum authorized by FDTD sampling to completely surround the area under test, i.e., the center of the CD). Limiting the number of probes (linearly located on a side of the CD) to 51 probes involves damaging TR focusing both in time (Figure 3.12c) and space (homogeneity of the E-field at focusing time is spoiled, as depicted in Figure 3.12d). The maximum level of focused E-field (e.g., E_x component) is decreased by more than a hundred scaling factors between the *ideal* TRC case (Figure 3.12a and b) and the TRM case (Figure 3.12c and d). In the following, particular care will be given to time focusing (Figure 3.12a and c) in comparison to space focusing (Figure 3.12c and d). To this end, various quantities are defined as follows:

– Maximum focusing signals (electromagnetic field/power and/or current/voltage). Given a particular time-reversed EM field E_{TR} (at location R_0 and time t), one may define the maximum focusing field by considering the useful part of the back-propagated signal (e.g., between −25 ns and 25 ns in Figure 3.12a and c) τ_u and defining the maximum TR signal at the space location R_0:

$$Max\left(R_0\right) = \max_{t\varepsilon\tau_u}(E_{TR}(R_0, t)) \tag{3.8}$$

– An important criterion dedicated to the assessment of the quality of the TR process is the signal-to-noise (STN) output. Theoretically, it was introduced in [29] requiring a definition of Heisenberg's time, as explained later (Section 3.3.2). It may also be computed by considering back-propagated and focused time signals (Figure 3.12a and c, for instance) as the ratio between the square of the magnitude of the focused peak

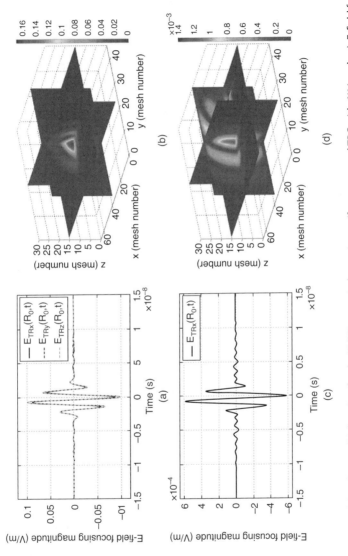

Figure 3.12 Time (a) and space (b) focusing for the EMTR numerical experiment (free space and TRC with 6114 probes). E-field focusing in time (c) and space (d) in the open space test case with TRM (54 probes).

over the time noise of E_{TR}^2 out of the duration of the useful signal (τ_u):

$$STN = \frac{E_{TR}(r = R_0, t = 0)^2}{E_{TR}^2(r = R_0, t \text{ not in } \tau_u)} \tag{3.9}$$

where $E_{TR}(R_0, t)$ stands for the reversed and focused field located at R_0 for time t. Similarly to relation (3.9), it is to be noted that signal-to-noise criterion also exists for the assessment of space focusing (but it will not be used in the following).

Based upon a previous report considering the huge requirements for a proper TR process in the open area, Yavuz and Teixeira proposed taking the benefit from multiple EM wave scattering in complex media in [30] using a complex bulk medium for random propagation. Relying on a previous test case for free space propagation (the same CD), the number of probes is restricted from the initial 51 to only 8 antennas. Whereas the maximum magnitude of an E-field reaches 100 mV/m in an open area test case with TRC (including more than 6000 probes), it decreases below 0.6 mV/m (E_x, the Cartesian component) with 51 TRM probes. The reverberation environment (i.e., relying on previous CDs with perfectly conducting boundary conditions) requires a limited number of probes (here eight antennas) with noticeable field enhancement at focusing time (around 50 mV/m at maximum for the E_y component, see Figure 3.13a) and space focusing homogeneity (Figure 3.13b). The trade-off between number of probes in TRM and levels of focusing in the EM field is improved in a reverberating environment. The following section will be focused on the most important parameters (duration of stimulation, number of probes, for instance) for the TR experiment in a reverberating environment.

3.3.2 Optimization of TR Parameters for the RC Numerical Experiment

The works from A. Derode *et al.* [31] demonstrated that the signal-to-noise (STN) ratio increases linearly with the square root of the number of sensors used (i.e., it means that each

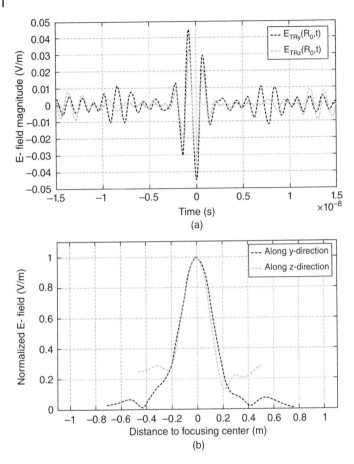

Figure 3.13 Time (a) and space (b) focusing for the EMTR numerical experiment (reverberating environment).

new sensor adds an amount of supplementary information theoretically non-correlated from known data). In reverberating environments (and particularly regarding the reverberating chamber, RC), elementary information is given by resonance modes; in order to illustrate the previous assumption, results in Figure 3.14 are obtained from FDTD simulations with the source parameter given by $fc = 400$ MHz, $\Delta\Omega = 260$ MHz, and varying number of sensors ($nc = 1$ to 20) randomly chosen inside the RC.

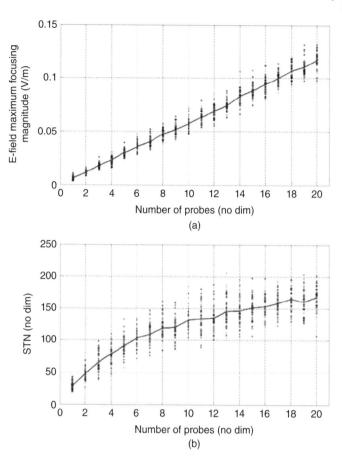

Figure 3.14 Importance of the number of sensors regarding the maximum focalization magnitude (a) and signal-to-noise ratio (b). Numerical experiments are achieved for 50 random locations of sources (markers) and trend (plain line) with $f_c = 400$ MHz and $\Delta\Omega = 260$ MHz.

Mean trends of reversed signal (i.e. maximum amplitude and STN) are computed from iteration of the whole process 50 times. Although Figure 3.14a clearly validates the linear increase of the focusing magnitude (maximum) with respect to the number of sensors, the use of STN criterion (Figure 3.14b) is telling more about the quality of the time-reversal (TR) process. Indeed, the maximum focusing magnitude is directly correlated to the sources (by their amplitude, location, etc.) whereas characterizing TR from STN offers more interactions with multiple reflections due to the reverberating environment. Thus, Figure 3.14b illustrates the increase of STN (square root evolution from 1 to 10 sensors) up to a balancing level where adding a complementary sensor does not improve STN anymore. The previous statement is easily understandable by considering the amount of available information that depends in the RC on the number of modes. The minimum number of sensors needed for optimized TR experiments are given by the ratio of Heisenberg's time over signal duration (e.g., in that case $\Delta H/\Delta t = 0.5~\mu s/0.065~\mu s \approx 8$, which is verified in Figure 3.14).

Thereafter, the impact of the signal duration considering STN of a reversed signal is investigated by considering TR experiments in RC with one unique sensor (TRM) and several durations of time-domain simulations (Δt with FDTD mock-up previously given).

Figure 3.15a shows that TR STN increases with frequency (mid-value of the frequency bandwidth) of initial time excitation; it is noted that stabilization is obtained after a certain duration ΔH called Heisenberg's time [32], which may be easily defined in a reverberating environment by counting the number of resonance modes. The numerical validation is obtained in Figure 3.15a with $\Delta H = 0.5~\mu s$, results expected from [32] (see this in the following relation (3.10)). The assumption of *information grains* was theoretically predicted in works from de Rosny *et al.* [33] for acoustics. Indeed, the pulse response of a reverberating enclosure is assumed to be linked with successive uncorrelated *information grains*, where the frequency bandwidth is about $1/\Delta t$. Figure 3.15b justifies the phenomenon of STN saturation depicted in Figure 3.15a since frequency responses given by longer simulation time duration (bottom line) are attached to an increase in the number of resonance modes. The number of *information grains* is deeply linked with the pulse duration (τ stands for the duration of the Gaussian sinus-modulated initial signal) and the total experiment duration Δt as follows: the information grains number is given by $\Delta t/\tau$ when $\Delta t << \Delta H$. For a longer time simulation ($\Delta t >> \Delta H$), the number of information grains is given by $1/(\tau^* \delta f)$, with δf the average gap between two

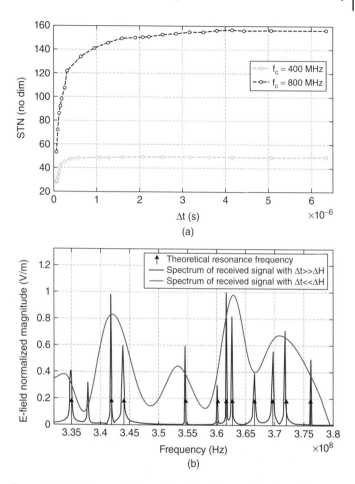

Figure 3.15 (a) TR impact of signal duration (Δt) regarding Heisenberg's time versus frequency. Two cases are considered here: $fc = 400$ MHz/$\Delta\Omega = 260$ MHz (case 1, black) and $fc = 800$ MHz/$\Delta\Omega = 260$ MHz (case 2, grey). (b) Influence of signal duration: information content versus theoretical resonance frequencies.

successive resonance modes from the cavity. Obviously, previous remarks justify the benefit that could be expected from the use of a classical experimental EMC device such as a reverberation chamber. In this context, the TR approach may be helpful for the radiated immunity testing process.

3.3.3 EMC Immunity Testing in a Reverberation Chamber

Based upon previous considerations regarding TR performances with respect to the number of sensors, signals and environment, a numerical illustration is proposed in Figure 3.16. The aim of this example is to validate for a practical test case the contribution of the TR procedure for EMC immunity testing. The equipment under test (EUT) is composed of metallic furniture (an opened cabinet) with different sublevels including three different dipoles acting as passive subsystems. The time input signal is a sine-Gaussian pattern from 100 MHz to 400 MHz (central frequency fc = 250 MHz with frequency bandwidth $\Delta\Omega$ = 400 MHz), and the time-reversal mirror (TRM) is composed by two half wavelength dipoles (L = 60 cm) located inside the reverberation chamber (Figure 3.1) of the Institut Pascal (IP, see experimental and numerical overviews, respectively, in Figures 3.1 and 3.16).

Proper choices both for the number of antennas and for the duration of the TR window are governed by Heisenberg's time in the IP MSRC, whose theoretical formulation is given by

$$\Delta H = 2\pi n(\omega) \tag{3.10}$$

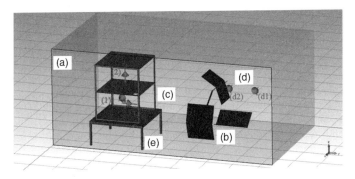

Figure 3.16 EMTR for EMC susceptibility testing; numerical mock-up (CST Microwave Studio®) including IP MSRC walls (a, transparency), paddles of mechanical stirrer (b), EUT (c, including locations of subsystems: dipoles 1, 2, and 3), TRM (d, two dipole antennas d1 and d2), and wooden holder (e, table).

where $n(\omega)$ is the modal density of the reverberation chamber, assumed to be constant over the whole frequency bandwidth $\Delta\Omega = 400$ MHz. Theoretically, $n(\omega)$ is well known for an ideal Faraday's cage (i.e., an empty one); in this case, some preliminary numerical tests were achieved (data are not shown here) including EUT and a mechanical stirrer. Heisenberg's time is extracted from the stabilization of the time-reversed STN rate; this duration is here about 10 µs. Such simulation duration implies too harsh numerical constraints (e.g. computing time) due to TR simulations in the IP MSRC (volume about 200 m³, number of configurations). For the previous reasons, a trade-off between the number of antennas available and time simulation, eight antennas are thus necessary to reach the stabilization of STN levels with time simulation durations of $\Delta t = 1.2$ µs. Since the number of antennas needed for TR experiments is given by the rate $\Delta H/\Delta t$, and assuming for computational reasons that $\Delta t = 4.25$ µs, two antennas were needed to achieve the next TR simulations.

This section is dedicated to the illustration of the TR capability to lead selective focusing regarding one of the three EUT components (Figures 3.17 and 3.18), whereas others are illuminated with weaker levels of electric (E) fields.

Assuming a test procedure where each of the three previous components requires the respect of various E-field thresholds (i.e. here the threshold for strict exposure limits of the equipment due to standards and/or safety reasons, for instance), we define, respectively, ET1 = 15 V/m, ET2 = 70 V/m, and ET3 = 40 V/m for each of the three components in Figure 3.16. After a first step recording impulse responses named $k_{ij}(t)$, with $1 \leq i \leq 2$ (antenna's label) and $1 \leq j \leq 3$ (system component number), and because of the assumed linearity of the whole system, it is possible to focus on any component (1, 2, or 3) with a chosen focusing magnitude of E-field via a straightforward post-treatment. Indeed, focusing on component 1 (Figures 3.17a and 3.18a) requires back-propagation of $pk_{11}(-t)$ and $pk_{21}(-t)$, respectively, via the first and second antennas. The coefficient p stands for the magnitude control offered by the TR process (for instance, it is possible to modify this parameter from the number of antennas, the duration time of simulation, and/or an external amplification system). Figures 3.17 and 3.18, respectively,

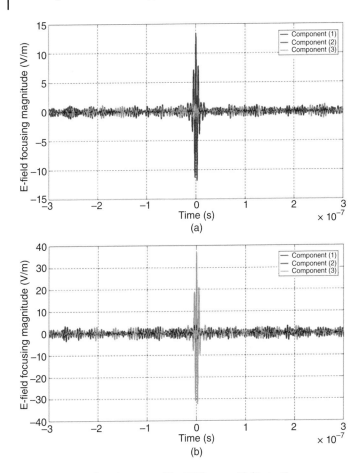

Figure 3.17 Time focusing control for EMC susceptibility testing:
(a) subsystem 1 (plain black line), $p = 1$; (b) subsystem 3 (grey plain line),
$p = 3$.

show the time (Figure 3.17) and space (Figure 3.18) focusing of
E-field for components 1 (Figure 3.17a and Figure 3.18a) and 3
(Figure 3.17b and Figure 3.18b). It should be noted that the "on
demand" focusing is respected (i.e., desired focusing is clearly
obtained on each of the subsystems). In Figure 3.18, space focusing represents the maximum E-field recorded over the whole
simulation duration and for each point of the sectional view.
It is to be noticed that, in each case, the maximum measured

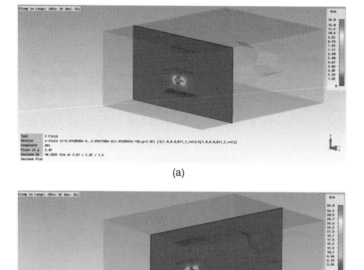

(a)

(b)

Figure 3.18 Space selective focusing for EMC susceptibility testing: (a) subsystem 1, $p = 1$; (b) subsystem 3, $p = 3$.

E-field is in good agreement with the considered component of the system. Moreover, the prescribed thresholds $ET1 = 15$ V/m and $ET3 = 40$ V/m are respected in this example (the same conclusion should be given also for component 2, although the data are not shown here).

Reaching the given level of E-field is not arbitrarily achieved. Indeed, the magnitude of the focusing field varies linearly with respect to the p-coefficient (data are not shown here). For instance, accessing the required threshold level of $ET1 = 15$ V/m needed no action ($p = 1$) since the maximum E-field over the simulation time was around 13 V/m. Even if the focusing magnitude should have been different from the required value, the linear combination with the p-parameter would have permitted the expected focusing level to be reached.

Following the previous idea, any combination is possible regarding the expected levels and/or the choice of the subsystems for focusing. Assuming different thresholds levels for components 1 and 3 (respectively ET1 = 13 V/m and ET3 = 35 V/m), linear weighted combinations of back-propagating signals allow selective focusing over one and/or two subsystem(s) at the same time. Practically, results in [32] validate the idea: the results are obtained (Figure 3.19a) by back-propagating

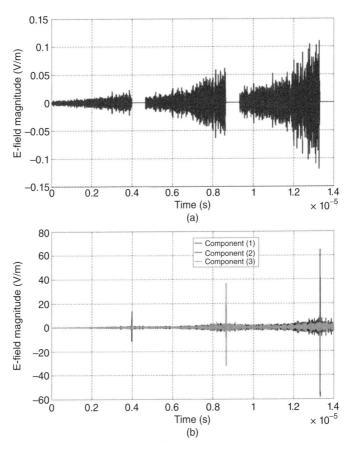

Figure 3.19 Focusing E-field at three different times over the three components. (a) Required signal back-propagated from antenna 1. (b) E-field with respect to time over each of the three components 1 (dark plain line), 2 (dash with circle line), and 3 (grey plain line) with focusing.

signals $p_1 k_{11}(-t) + p_3 k_{13}(-t)$ with $p_1 = 1$ and $p_3 = 3$ with the first antenna (Figure 3.19a) and $p_1 k_{21}(-t) + p_3 k_{23}(-t)$ with the second antenna. The same principle may be experienced to provide focusing on each component at different times by playing with the back-propagating time. Thus, by providing signal $p_1 k_{11}(-t) + p_2 k_{12}(-t + t_2) + p_3 k_{13}(-t + t_1)$ to antenna 1 and signal $p_1 k_{21}(-t) + p_2 k_{22}(-t + t_2) + p_3 k_{23}(-t + t_1)$ to antenna 2 (with t_1 and t_2 times standing for the expected time shifts, respectively, for focusing on components 3 and 2), one may obtain a focusing over the different subsystems at different times, but respectful of the required given E-field threshold (Figure 3.19b). This offers the possibility to achieve pulse generation with periodical time and may be useful in various domains, including EMC standard test procedures as an alternative to continuous wave (CW) excitations [34].

3.4 Towards Realistic EMC Testing

Based upon works in [34] and [35], TR methodology requires a particular statistical care to outputs. Indeed, the assessment of the TE quality is needed in particular regarding the signal-to-noise (STN) ratio for focused fields. Another important aspect of the whole process relies on the existence of uncertainties since certain parameters are not completely controlled: it should be noted that locations of antennas and material electrical and geometrical properties may be partially known, even purely assumed due to the complexity of EMTR (e.g., taking into account, for instance, measurement restrictions in terms of signal generation quality). The aim of this section is to demonstrate the ability of the sensitivity analysis technique jointly with stochastic methods to truthfully assess the TR robustness.

3.4.1 Practical Limitations

Obviously, practical investigations involving TR purposes are subject to several sources of limitations. Naturally, one may think about particular requirements due to materials and/or the setup itself (complete knowledge of parameters such as sensor location, electric properties of materials, for instance). One of the

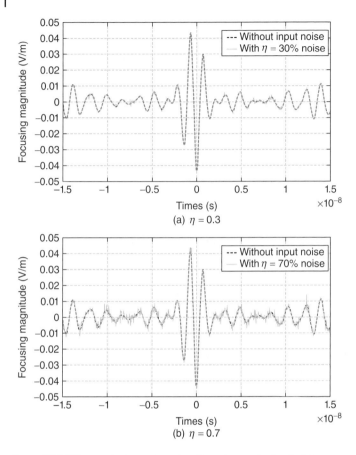

Figure 3.20 TR numerical experiments with various noise levels: (a) $\eta =$ 30% and (b) $\eta =$ 70% (dashed line) and without input noise (plain line).

very first worries about TR was its robustness to the insertion of experimental noise. To that end, the same computational case from Subsection 3.3.1 is adapted in the following.

Figure 3.20 gives an overview of "numerical" robustness of the TR process subjected to noise. In this context, Gaussian noise (characterized by the mean value given by the initial signal magnitude and coefficient of variation η, i.e., the ratio of standard deviation over mean) is added to the initial signal received by TR sensors. Indeed, even with a high coefficient of variation $\eta = 0.7$

(Figure 3.20b), both maximum focusing magnitude and STN ratios are comparable with results obtained without noise (plain line). These very first results are purely qualitative ones; despite all, they focus on the relative robustness of the TR procedure. In the following, a whole methodology to enrich the quantification of TR robustness by means of statistical outputs is proposed.

In order to assess the relative statistical robustness of the TR process in a reverberating environment, we put the focus on previous works. In this context, some works were produced to demonstrate the capability of TR to provide enhancement of field control for several applications including defense [36] (Figure 3.21a) and EMC [37, 38].

3.4.2   Robustness and Statistics

This part will place emphasis on simulations achieved in a reverberation chamber (RC) from the Institut Pascal (IP) enclosed with a cabinet designed [39] from scratch (the inner view is given in Figure 3.21b). As previously validated, enclosed environments (and RC especially) prove to be particularly interesting both by preventing any external electromagnetic disturbances and enhancing TR performances. The goal of manufacturing the device shown in Figure 3.21b was to control precisely several typical parameters: slot, sizes of enclosures, location of subsystems. The governing idea was to propose a system related to EMC standard concerns (e.g., for estimating the shielding effectiveness (SE) of enclosures [40]) and statistical purposes. Moreover, the idea was also to propose a device close to the cabinet developed in [28] for experimental field enhancement purposes, as depicted in Figure 3.21a. Deeply linked to typical EMI/EMC concerns, a crucial parameter in many applications dealing with EM field penetration relies on apertures. In this EMC framework, TR provides many advantages but its efficiency theoretically requires infinite knowledge of propagative parameters (antenna locations, bulk medium, for instance). The experimental development of such a device was part of a program funded by the Armaments Procurement Agency (DGA in French) of the French Ministry of Defense called *REI "PRINCE"* and dealing with the integration of uncertainties for EMC issues [39]. The

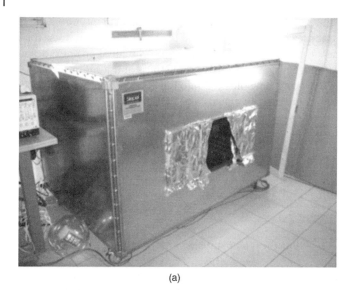

(a)

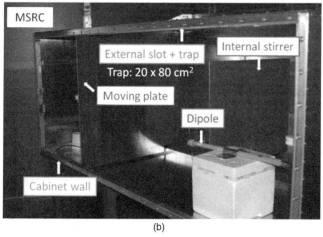

(b)

Figure 3.21 Experimental setups for (a) experimental enhancement of EMTR [28] and (b) statistical properties of fields [39].

aim of this program was to demonstrate the interest of stochastic techniques in an EMC framework.

The proposed test case (Figure 3.21b) aims at assessing the robustness of TR regarding intentional or non-intentional events during the TR process. Indeed, different variations may have an impact on the quality of EMTR focusing: geometrical ones (displacements/deformations of scatterers and components due to environment drifts and/or a human experimenter), material ones (propagative bulk medium, electrical properties of devices), the measurement process (uncertainties due to repeatability, locations of sources, and/or probes, for instance). In the following, each of these sources of uncertainty will be considered as independent random variables (RVs) for the sake of simplicity (solutions exist if dependence of parameters is required). Particular care will be taken to variations happening between the two TR steps (i.e., between emission/recording and back-propagation of time-reversed signals) since this case should maximize spoiling effects due to uncertainties.

3.4.2.1   Sensitivity Analysis for EMTR

In the early 1990s, an approach was proposed by Morris [41] to identify the "inputs" of a given modeling method (for instance, throughout arbitrary mapping $f(X)$ stood, for instance, for the maximum E-field focusing in the function of random vector X). The work aimed at determining which "inputs" may lead to important effects on a given "output". Nowadays, the model is widely used since it is easy to compute and is particularly efficient with the problems containing hundreds of "inputs". Therefore, among the numerous methods available to conduct SA, in this research the method introduced by Morris is used [41] (see [42] for a complementary remark about SA). Morris' method (MM) is a "screening" sensitivity analysis technique. As aforementioned, MM enables a quick assessment of the variations of the model's outputs in relation to input variations for a large number of the model's parameters. The key idea is to sample inputs over their initial range. The sequence for MM may be summarized by assuming a mapping with d inputs and a single output: $X = (x_1, x_2, \ldots, x_d) \mapsto f(X)$; for instance, let us assume X is a random vector representing the uncertain parameters during the TR process (locations of antennas, material properties, etc.). The key idea is to sample the input random spaces and assess the effect of a weak variation over one variable at a time (OAT). This variation is computed throughout an elementary influence criterion obtained from mapping $f(X)$: $e_k(X_i)$, with X_i the random parameter number i for the kth assessment of the influence criterion. The process is reproduced r times ($k = 1,\ldots, r$) up to RV X_d. The mean $\mu^*(X_i)$ and standard deviation $\sigma(X_i)$ of the elementary effects for each RV are given by

$$\mu^*(X_i) = \frac{1}{r} \sum_{k=1}^{r} |e_k(X_i)| \tag{3.11}$$

$$\sigma(X_i) = \sqrt{\frac{1}{r} \sum_{k=1}^{r} e_k(X_i)^2 - \mu^*(X_i)^2} \tag{3.12}$$

Obviously, relations (3.11) and (3.12) are significant when modeling the impact of RVs:

- μ^* weak and σ weak $\Rightarrow$ marginal influence
- μ^* high and σ weak $\Rightarrow$ linear influence, almost zero interaction
- μ^* high and σ high $\Rightarrow$ non-linear impact and/or interactions.

Interpretation of the results is obtained from the chart $\sigma = g(\mu^*)$, as depicted later in Figure 3.23. In a first step, the most influential parameters are given on the right side of the chart. In a second step, it is also interesting to rank RVs with interactions (upper side of the chart) from RVs with linear effects located close to the abscissa axis. The interested reader may find plenty of references regarding use of MM in various domains including economics, mechanics, applied mathematics, and electrical engineering (see, for instance, [42]).

As aforementioned, this part is now dedicated to modeling the TR numerical experiment under uncertain assumptions regarding various kinds of parameters including geometrical variations of the environment (sizes/locations of scatterers, positions of probes). The numerical case under investigation refers to modeling the Institut Pascal MSRC (Figures 3.1 and 3.16). In the presence of a mechanical mode stirrer, the cabinet depicted in Figure 3.21b is modeled jointly with two perfectly conducting cubes located in the vicinity of the external aperture and of two cubic scatterers (Figure 3.22).

The numerical setup of the TR experiment chosen in the following is depicted in Figure 3.22. It is inspired from experiments [39] achieved in the Institut Pascal MSRC aiming at demonstrating the interest of statistical approaches for shielding

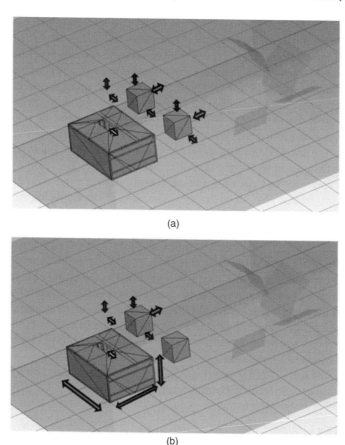

(a)

(b)

Figure 3.22 Numerical EMTR experiments (inside the IP MSRC): (a) test case 1 (moving scatterers out of the cabinet, size of aperture, and location of dipole source, see Figure 3.21b) and (b) test case 2 (varying sizes of the cabinet, size of aperture, and location of dipole source, see Figure 3.21b).

effectiveness of the cabinet. The enclosure is settled in the working volume (WV) of the RC whereas the TRM is composed of a unique dipole source S (above the enclosure in Figure 3.22) located out of the cabinet and matched on 50 Ω impedance. Similarly to the electromagnetic source S, a dipole receiver R is placed inside the enclosure (masked in Figure 3.22). In the following, we will focus on the current I_R collected on receiver R (initial 1 V excitation from source S, with a sine Gaussian pattern: central frequency f_c = 2 GHz, $\Delta\Omega$ = 2 GHz). A particular emphasis is given respectively regarding the maximum (relative to the simulation time) focusing magnitude (in µA) and the signal-to-noise (STN) rate of current I_R. Based upon the deterministic test case, the EMTR process is achieved and the evolution of current I_R over time is recorded after the second TR step. It is assumed in the following that some parameters may be subject to uncertain variations (especially between the two TR steps: emission and back-propagation of fields): the location of source S and receiver R, positions of scatterers, width of external aperture, and/or sizes of enclosures. The aims of the following two subsections will be split into two main steps: first, identify the most influential parameters; then, quantify the effect of random variations in view of the deterministic ("ideal") case (i.e., without uncertain assumptions). In this context, we will put the focus on the so-called magnitude rate (MR), defined as the ratio between the level of maximum focusing current obtained over receiver R in the deterministic case and the mean value given with uncertain assumptions. The definitions of random parameters (distributions and second-order statistics) are given in Tables 3.1 and 3.2 for test cases 1 and 2, respectively.

The second-order statistics (i.e., means and coefficient of variations, CV) and distributions of all random parameters are given

Table 3.1 Second-order statistics and distributions of random parameters relative to test case 1.

RV number	RV	Distribution	Mean (cm)	Coefficient of variation (%)
1	SxS	Uniform	325	0.18
2	SzS	Uniform	250	0.23
3	OpnA	Uniform	279.5	2.07
4	SxC1	Uniform	325	0.18
5	SyC1	Uniform	300	0.19
6	SzC1	Uniform	200	0.29
7	SxC2	Uniform	500	0.12
8	SyC2	Uniform	300	0.19
9	SzC2	Uniform	200	0.29

Table 3.2 Second-order statistics and distributions of random parameters relative to test case 2.

RV number	RV	Distribution	Mean (cm)	Coefficient of variation (%)
1	SxS	Uniform	325	0.18
2	SzS	Uniform	250	0.23
3	OpnA	Uniform	279.5	2.07
4	SxC1	Uniform	325	0.18
5	SyC1	Uniform	300	0.19
6	SzC1	Uniform	200	0.29
7	VarCabX	Uniform	195.5	0.30
8	VarCabY	Uniform	140	0.41
9	VarCabZ	Uniform	76	0.76

in Table 3.1 for test case 1. In that framework, nine RVs are considered regarding: location of the source antenna, respectively, in the x- and z-directions (SxS and SzS; see Figure 3.22a); width of the aperture (external to the enclosure, OpnA; see Figure 3.22a); and positions (centers) of cubic scatterers in each of the three Cartesian components (SxC1, SyC1, SzC1, and SxC2, SyC2, SzC2; see Figure 3.22a). The following results were obtained with the finite integral technique (FIT) time-domain models (CST MWS®). The total duration of each simulation is $Tf = 560$ ns. Computing was performed with an Intel Xeon processor at 3.30 GHz and with 16 Go RAM, and it needed around 10 minutes per simulation.

Similarly to the previous case, Table 3.2 synthesizes the random assumptions for test case 2, where uncertain parameters are still focused on the location of the source antenna (SxS and SzS), on the width of the aperture (OpnA), and on the position of the first cubic scatterer (SxC1, SyC1, SzC1). Differences rely on assuming random variations around the sizes of the enclosure, which was experimentally observed with the cabinet in Figure 3.21b regarding the duration of experiments (days) and intrinsic drifts of the ambient temperature (during days).

The simulations required by MM [42] are given in Figure 3.23. As previously explained, the charts $\sigma = g(\mu^*)$ give an overview of the most influential parameters (circled in Figure 3.23). Thus,

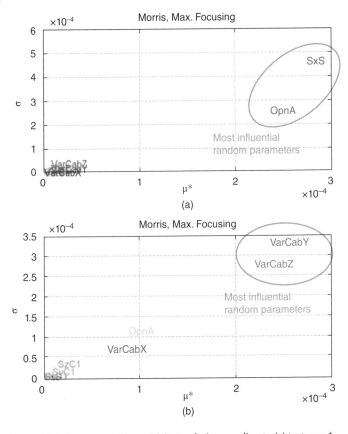

Figure 3.23 Screening MM sensitivity analysis according to (a) test case 1 and (b) test case 2 for TR maximum focusing. Screening MM sensitivity analysis according to (c) test case 1 and (d) test case 2 for the TR STN ratio. (*Continued on next page.*)

Morris' figures of importance lay emphasis on variables SxS (drifts along the x-direction for source S) and the width of external aperture both considering the maximum focusing current I_R (Figure 3.23a) and the STN ratio. It is to be noted that the second test case (2) shows a different ranking of influential parameters: two RVs seem to play a key role here regarding the sizes of the cabinet along the y- and z-directions. Considering the reverberating nature of the proposed study (field penetration

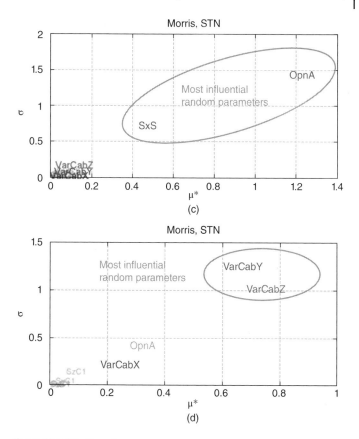

Figure 3.23 (*Continued*)

inside a cabinet settled inside the RC), the importance of random variations around the size of the cabinet was obviously expected. Screening interpretation via MM provides a confirmation of a previous remark and allows the focus to be put on the random description of different parameters regarding test cases 1 and 2. Indeed, in the following, we will only consider two random parameters: SxS and OpnA in test case 1 and VarCabY and VarCabZ in test case 2. This substantially simplifies the statistical study of the properties of the TR process, as explained in the next subsection.

3.4.2.2 Statistical Robustness of the TR Process

This section is devoted to the statistical assessment of EMTR focusing outputs (e.g., maximum focusing field, STN ratio, etc.). Far from giving a complete view of statistical and probabilistic techniques that could be used in that context, this section aims at providing sufficient theoretical elements to understand better the philosophy of the proposed methodology. More about the state of the art of spectral and statistical methods can be found in [43] and [44]. Roughly speaking, the main target of sampling techniques is to provide a competitive experimental design of chosen weighted points (often called "sigma" points) to describe the space of input random parameters. Similarly to Monte Carlo (MC) techniques with random sampling, an experimental design is expected to be cost effective (i.e., in terms of the number of points) and should remain accurate enough to properly describe mapping variations (for instance here the maximum focusing field obtained from *full-wave* simulations of the TR process). Without exhaustiveness, the following will lay emphasis on the stochastic collocation method (SCM), an efficient technique based upon polynomial expansion of required statistics (e.g., the mean and variance of the maximum focusing field). Key theoretical parameters of the SCM and some examples of its application in electrical engineering may be found in [39] and [45]. Despite all, to illustrate its use in this research, hereafter we will present the philosophy of the technique step by step assuming only one RV. Without any lack of generality, use of the SCM may be extended to multi-RV issues.

Let us assume a mapping given through function $f{:}R \rightarrow R$ and an RV X given by an a priori probabilistic distribution. The goal of this part is to establish the relations needed to access the mean and variance of RV $Y = f(X)$. A polynomial expression relying on the use of Lagrange polynomials is proposed with Gauss quadrature integration rules; this leads to

$$E[f(X)] \approx \sum_{i=0}^{n} \omega_i f(x_i) \tag{3.13}$$

$$var(f(X)) \approx \sum_{i=0}^{n} \omega_i (f(x_i))^2 - E[f(X)]^2 \tag{3.14}$$

with (ω_i, x_i) a set of weighted points x_i needed to compute the mean $E[f(X)]$ and variance $var(f(X))$ of mapping $f(\cdot)$ (for instance the average of the maximum focusing current over a dipole antenna here). Generally, the SCM requires less than 10 calls (i.e., $n = 2, \dots, 8$ in relation (3.14)) to deterministic mapping (numerical code, experimental set of measurements, etc.) per random variable.

In the following, the first-order (mean) statistics of the maximum focusing field are assessed for test cases 1 and 2 from the SCM with an increasing order expansion (i.e., requiring an increasing number of calls to *full-wave* simulations). Those results will be compared to the mean value computed using the MC technique in order to provide a "reference" result (here 500 simulations were achieved). Finally, the statistical data are normalized and faced with the deterministic "ideal" case (i.e., without uncertain assumptions following Tables 3.1 and 3.2) in order to quantify the sensitivity of the TR process with regard to uncertain parameters.

Figure 3.24 shows the statistical convergence of TR outputs: maximum focusing magnitude of current (normalized with deterministic case) for test case 1 (Figure 3.24a) and test case 2 (Figure 3.24b). It should be noted that the MC convergence is assessed from an initial set of 500 random samples (i.e., from 500 simulations with random generation of inputs given in Tables 3.1 and 3.2). In this framework, it is expected that sizes of the enclosure will play a crucial role since resonances are intrinsically dependent on RVs VarCabX, VarCabY and VarCabZ (text case 2). The previous assumption was validated considering Morris' analysis in Figure 3.23b and d.

Test case 1 (Figure 3.24a) is characterized by a significant convergence of the MC mean of maximum focusing current magnitude for the TR process subject to uncertain assumptions. The convergence of SCM is obtained by considering increasing polynomial expansion orders, respectively, with 3×3, 5×5, 7×7, and 9×9 points in the experimental design set of points (see Figure 3.24a). Even with the highest order (i.e., with $9 \times 9 = 81$ points), good agreement is noticed with MC data. Moreover, the computing speedup of the combined use of MM jointly with SCM is around 48% relative to the MC approach (500 simulations). Indeed, in that framework (i.e., test case 1), 100 realizations are necessary to achieve Morris' treatments (Figure 3.23a and c), exhibiting the major role of RVs linked to source

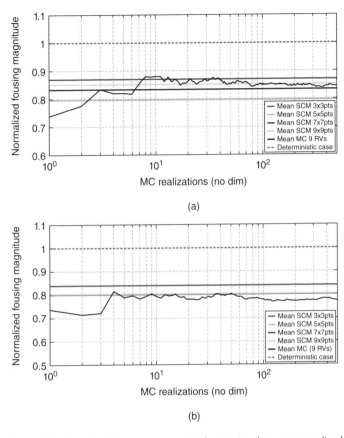

Figure 3.24 Statistical TR outputs (magnitude rate; i.e., data are normalized to the focusing magnitude obtained without uncertain assumptions) relative to (a) test case 1 and (b) test case 2 for TR maximum focusing.

x-location (SxS) and aperture opening (OpnA). Then, expanding orders with at least nine points per RV is necessary in order to lead to $100 + 161 = 261$ calls to the *full-wave* simulating tool. Of course, the number of MC simulations was arbitrarily chosen here (500 simulations to provide realistic durations of simulation), enhancing the level of convergence needed for the MC strategy to increase the competitiveness of the SCM.

Similarly to the previous case, Figure 3.24b shows the quick convergence of the SCM since only 25 points seem sufficient to

Table 3.3 Statistical assessment of TR outputs for test cases 1 and 2.

Output	MC	SCM	Deterministic	CV	MR
Maximum focusing magnitude (case 1)	2980 μA	3000 μA	3519 μA	16.4 %	85.2 %
STN (no dimensions case 1)	11.34	11.67	14.54	19.0 %	80.3 %
Maximum focusing magnitude (case 2)	2719 μA	2803 μA	3493 μA	15.8 %	80.3 %
STN (no dimensions case 2)	11.4	12.35	14.47	20.1 %	85.4 %

reach converged data from hundreds of MC simulations including 9 RVs. Consequently, the overall cost from joint use of MM and SCM is valuable since only 133 simulations (100 for MM and $9 + 25 - 1$ for SCM expansions, respectively, with 3×3 and 5×5 points) are required compared to 500 MC realizations needed in this example. The benefit in terms of computing costs that should be expected from the SCM is at least about 74% speedup in comparison to the brute MC strategy. Indeed, by questioning MC convergence, the gain expected from MM and SCM use in this case may be far greater than 75%.

Close to results in Figure 3.24, Table 3.3 provides data regarding test cases 1 and 2, both for the maximum focusing current and STN ratio. Moreover, each of the mean results (given here in absolute with magnitude in μA) obtained from previous TR outputs are enriched with the second-order statistics (i.e., the coefficient of variation, CV, standing for the ratio between the standard deviation and average value).

First, the information about the sensitivity of the TR process to random assumptions is valuable since the CV is found to be between 16% and 20%. Moreover, it is important to notice that in both cases (i.e., 1 and 2), the given methodology provides an assessment of TR spoiling due to uncertain assumptions throughout the MR coefficient. It is to be noted that the TR procedure is relatively saved regarding random variations depicted by RVs in Tables 3.1 and 3.2 since, even for STN output, MR is greater than 80%.

3.5 Conclusion

This chapter demonstrated the interest of EMTR in an EMC framework. First, the important issue of cables, transmission lines, and power networks in EMC can benefit from the use of EMTR properties. Wire diagnosis, especially the detection, localization, and characterization of soft defect, or soft fault, is efficiently and accurately addressed by various time-reversal processes. Theoretical and numerical contents proposed in this chapter have illustrated this potential application of TR, which has been recently experimentally confirmed even though Telegrapher's equations are not strictly time-reversal invariant [10, 46].

EMTR appeared as an accurate and efficient procedure in the wire diagnosis context. It is also a relevant tool for EMC radiated immunity testing. Relying on the foundations of TR, it has been underlined that EMTR may offer great benefits for EMC standard tests, especially considering the use of reverberation chambers. Of course, parametric studies are required to properly define the characteristics of the EMTR experimental setup (number of antennas, duration of stimulations, location, and scattering environment) in accordance with the required frequency bandwidth. A previous point was validated by considering various enclosures (including the Institut Pascal's own MSRC), based upon numerical time-domain tools (finite difference and finite integral techniques) and theoretical EMTR expectations. Selective focusing allowed by EMTR was achieved from simulations of the Institut Pascal's own MSRC. It authorized the use to stress part of generic equipment under test (EUT) with arbitrarily chosen levels of electric (E) fields. It is to be noted that a large benefit could be taken from use of the EMTR procedure, since the rest of the computational domain (including a potentially sensitive part of the EUT) bore very weak levels of E-field in comparison to the focused one. Moreover, time shape, polarization, even incidence of the radiated test signal may be a priori chosen.

Although previous use of EMTR in a reverberating environment was validated in the EMC context, it is well known that the entire TR procedure might be sensitive to any kind of variation due to the nature of EMC sources and victims (e.g., aging, non-stationarity, location of antennas), and/or the coupling paths involved (materials, geometry of scatterers, shielding, environmental variations including moisture, thermal/mechanical conditions, for instance). Theoretically, EMTR is subject to ideal assumptions considering knowledge of the geometrical, mechanical, thermal, and electromagnetic conditions where TR experiments are achieved. The aim of this chapter was also to quantify TR robustness when assuming uncertain variations around input parameters including locations of sensors/antennas and sizes of scatterers. A statistical study of TR intrinsic characteristics was proposed (including the maximum focused field and signal-to-noise ratio). Due to the natural complexity of the TR process, a first step was necessary to determine the most influential parameters (sensitivity analysis, SA). The stochastic work was achieved by considering empirical variations of random parameters and using the stochastic collocation method (SCM) to minimize the number of simulations needed for statistics. Results obtained from joint approaches including SA and SCM were in agreement with reference Monte Carlo data (involving the overall inputs). In this context, EMTR spoiling due to random variations was quantified throughout the decrease of the maximum focused field and STN ratio. Since the proposed methodology is entirely non-intrusive, its applicability to experimental measurements is ensured.

Relying on the noticeable properties of EMTR, EMC test procedures might hugely benefit from application of the TR basis as depicted in this chapter. Many works are nowadays in progress to enhance EMTR capabilities jointly with EMC concerns.

References

1 Directive 2014/30/EU of the European Parliament and of the Council of 26 February 2014 on the "Harmonisation of the laws of the Member States relating to electromagnetic compatibility," *Official Journal of the European Union*, 29.3.2014.

2 P. Pajusco and P. Pagani, "On the use of uniform circular arrays for characterizing UWB time reversal," *IEEE Transactions on Antennas and Propagation*, vol. 57, no. 1, pp. 102–109, 2009.

3 C. R. Paul, *Analysis of Multi-conductor Transmission Lines*. New York: John Wiley & Sons, Inc., 1994.

4 N. V. Kantartzis and T. D. Tsiboukis, *Modern EMC Analysis Techniques: Volume I: Time-Domain Computational Schemes and Volume II: Models and Applications*. Morgan & Claypool, 2008.

5 C. R. Paul, *Introduction to Electromagnetic Compatibility*, 2nd edition, ISBN: 978-0-471-75814-3. New York: John Wiley & Sons, Inc., 2006.

6 P. Besnier and B. Démoulin, *Electromagnetic Reverberation Chambers*, ISBN: 978-1-84821-293-0, August 2011. Wiley-ISTE, 2011.

7 A. Taflove and S. C. Hagness, *Computational Electrodynamics: The Finite-Difference Time-Domain Method*, 3rd edition. Artech House, 2005.

8 F. Auzanneau, "Wire troubleshooting and diagnosis: review and perspectives," *Progress in Electromagnetics Research B*, vol. 49, pp. 253–279, 2013.

9 M. Kafal, A. Cozza, and L. Pichon, "Locating faults with high resolution using single-frequency TR-MUSIC processing," *IEEE Transactions on Instrumentation and Measurement*, pp. 1–7, early access June 2016.

10 L. Sahmarany, L. Berry, N. Ravot, F. Auzanneau, and P. Bonnet, "Time reversal for soft faults diagnosis in wire networks," *Progress In Electromagnetics Research M*, vol. 31, pp. 45–58, 2013.

11 L. Sahmarany, L. Berry, K. Kerroum, F. Auzanneau, and P. Bonnet, "Time-reversal for wiring diagnosis," *Electronics Letters*, vol. 48, no. 21, pp. 1343–1344, October 2012.

12 L. Abboud, A. Cozza, and L. Pichon, "Utilization of matched pulses to improve fault detection in wire networks," in *Proceedings of the 9th International Conference on ITS Telecomm*, Lille, France, 2009.

13 G. Lerosey, J. de Rosny, A. Tourin, A. Derode, and M. Fink, "Time reversal of electromagnetic waves," *Physical Review Letters*, vol. 92, 193904 pp. 1–3, 2004.

14 A. Derode, A. Tourin, J. de Rosny, M. Tanter, S. Yon, and M. Fink, "Taking advantage of multiple scattering to communicate with time-reversal antennas," *Physical Review Letters*, vol. 90, no. 1, 014301 pp. 1–4, 2003.

15 L. Abboud, A. Cozza, and L. Pichon, "A matched-pulse approach for soft-fault detection in complex wire networks," *IEEE Transactions on Instrumentation and Measurement, Institute of Electrical and Electronics Engineers*, vol. 61, no. 6, pp. 1719–1732, 2012.

16 M. Fink, C. Prada, F. Wu, and D. Cassereau, "Self focusing in inhomogeneous media with time reversal acoustic mirrors," in *Proceedings of the IEEE Ultrasonics Symposium*, pp. 681–686, 1989.

17 C. Prada, S. Manneville, D. Spoliansky, and M. Fink, "Decomposition of the time reversal operator: detection and selective focusing on two scatterers," *The Journal of the Acoustical Society of America*, vol. 99, no. 4, pp. 2067–2076, 1996.

18 H. Lev-Ari and A. Devancy, "The time-reversal technique re-interpreted: subspace-based signal processing for multi-static target location," in *Proceedings of the 2000 IEEE Workshop on Sensor Array and Multichannel Signal Processing*, pp. 509–513, 2000.

19 M. Yavuz and F. Teixeira, "Full time-domain DORT for ultrawideband electromagnetic fields in dispersive, random inhomogeneous media," *IEEE Transactions on Antennas and Propagation*, vol. 54, no. 8, pp. 2305–2315, 2006.

20 H. Tortel, G. Micolau, and M. Saillard, M. "Decomposition of the time reversal operator for electromagnetic scattering," *Journal of Electromagnetic Waves and Applications*, vol. 13, no. 5, pp. 687–719, 1999.

21 G. Micolau and M. Saillard, "D.O.R.T. method as applied to electromagnetic subsurface sensing," *Radio Science*, vol. 38, no. 3, pp. 1–12, 2003.

22 L. Abboud, A. Cozza, and L. Pichon, "A non-iterative method for locating soft faults in complex wire networks," *IEEE Transactions on Vehicular Technology*, vol. 62, no. 3, pp. 1010–1019, 2013.

23 M. Kafal, A. Cozza, and L. Pichon, "Locating multiple soft faults in wire networks using an alternative DORT implementation," *IEEE Transactions on Instrumentation and Measurement*, vol. 65, no. 2, pp. 399–406, 2015.

24 M. Kafal, A. Cozza, and L. Pichon, "An efficient technique based on DORT method to locate multiple soft faults in wiring network," in *Proceedings of the IEEE AUTOTESTCON*, National Harbor, MD, pp. 339–344, 2015.

25 H. Manesh, G. Lugrin, R. Razzaghi, C. Romero, M. Paolone, and F. Rachidi, "A new method to locate faults in power networks based on electromagnetic time reversal," in *Proceedings of the IEEE 13th International Workshop on Signal Processing Advances in Wireless Communications*, 2012.

26 R. Razzaghi, G. Lugrin, H. Manesh, C. Romero, M. Paolone, and F. Rachidi, "An efficient method based on the electromagnetic time reversal to locate faults in power networks," *IEEE Transactions on Power Delivery*, vol. 28, pp. 1663–1673, 2013.

27 L. Sahmarany, F. Auzanneau, and P. Bonnet, "Novel reflectometry method based on time reversal for cable aging characterization," in *58th IEEE Holm Conference on Electrical Contacts*, 23–26 September 2012, Portland, Oregon, 2012.

28 M. Davy, "Application du retournement temporel en micro-ondes à l'amplification d'impulsions et l'imagerie" (in French), PhD thesis, Université Paris 7, October 2010.

29 J. de Rosny, "Milieux réverbérants et réversibilité" (in French), PhD thesis, Université Paris 6, 2000.

30 M. E. Yavuz and F. L. Teixeira, "A numerical study of time-reversed UWB electromagnetic waves in continuous random media," *IEEE Transactions on AWPL*, vol. 4, pp. 43–46, 2005.

31 A. Derode, A. Tourin, and M. Fink, "Limits of time-reversal focusing through multiple scattering: long-range correlation," *Journal of the Acoustical Society of America*, vol. 107, no. 6, pp. 2987–2998, 2000.

32 I. El Baba, S. Lalléchère, and P. Bonnet, *Time Reversal for Electromagnetism: Applications in Electromagnetic Compatibility, Trends in Electromagnetism – From Fundamentals to Applications*, Intech, ISBN 978-953-51-0267-0, 2012.

33 J. de Rosny and M. Fink, "Overcoming the diffraction limit in wave physics using a time-reversal mirror and a novel acoustic sink," *Physical Review Letters*, vol. 89, no. 12, 124301 pp. 1–4, 2002.

34 H. Vallon, A. Cozza, F. Monsef, and A. S. Chauchat, "Time-reversed excitation of reverberation chambers:

improving efficiency and reliability in the generation of radiated stress," *IEEE Transactions on EMC*, vol. 58, no. 2, pp. 364–370, 2016.

35 A. Cozza, "Statistics of the performance of time reversal in a lossy reverberating medium," *Physical Reviews E*, vol. 80, no. 5, 056604 pp. 1–11, 2009.

36 M. Davy, J. de Rosny, J. C. Joly, and M., Fink, "Focusing and amplification of electromagnetic waves by time-reversal in a leaky reverberation chamber," *CRAS*, vol. 11, no. 1, 2010.

37 A. Cozza and H. Moussa, "Enforcing deterministic polarisation in a reverberationg environment, *Electronics Letters*, vol. 45, no. 25, pp. 1299–1301, 2009.

38 A. Cozza, "Increasing peak-field generation efficiency of reverberation chamber," *Electronics Letters*, vol. 46, no. 1, pp. 38–39, 2010.

39 S. Lalléchère, S. Girard, P. Bonnet, and F. Paladian, Stochastic approaches for ElectroMagnetic Compatibility: a paradigm from complex reverberating enclosures, in *Proceedings of the ESA Workshop on EMC*, Venice, Italy, 2012.

40 International Electrotechnical Commission (IEC), IEC 61000-4-21, EMC - Part 4-21: Testing and Measurement Techniques – Reverberation Chamber Test Methods, 2003.

41 M. D. Morris, "Factorial sampling plans for preliminary computational experiments," *Technometrics*, vol. 33, no. 2, pp. 161–174, 1991.

42 A. Saltelli, M. Ratto, S. Tarantola, and F. Campolongo, "Sensitivity analysis practices: strategies for model-based inference," *Reliability Engineering System Safety*, vol. 91, no. 10–11, pp. 1109–1125, 2006.

43 D. Xiu, "Fast numerical methods for stochastic computations: a review," *Communication and Computation Physics*, vol. 5, no. 2, pp. 242–272, February 2009.

44 D. P. Kroese, T. Taimre, and Z. I. Botev, *Handbook of Monte Carlo Methods*, Wiley Series in Probability and Statistics, 2011.

45 P. Bonnet, F. Diouf, C. Chauvière, S., Lalléchère, M. Fogli, and F. Paladian, "Numerical simulation of a reverberation chamber with a stochastic collocation method," *CRAS*, vol. 10, no. 1, pp. 54–64, 2009.

46 G. Lugrin, R. Razzaghi, F. Rachidi, and M. Paolone, "Electromagnetic time reversal applied to fault detection: the issue of losses," in *2015 IEEE International Symposium on Electromagnetic Compatibility (EMC)*, Dresden, pp. 209–212, 2015.

4

Amplification of Electromagnetic Waves Using Time Reversal

Matthieu Davy,[1] Mathias Fink,[2] and Julien de Rosny[2]

[1] Institut d'Electronique et de Télécommunications de Rennes, University of Rennes 1, France
[2] Institut Langevin, ESPCI ParisTech, Paris, France

4.1 Outline

Traditional focusing systems of wideband signals make use of a beamforming method applied to an array of antennas. The signals emitted by the antennas are time-delayed so that they interfere constructively at the focal point. The amplitude of the focused signal is then proportional to the sum of the amplitudes of the short pulses generated by each antenna. In this chapter, we present a different approach based on the time-reversal technique to focus high-amplitude wideband pulses. We take advantage of a leaky reverberating chamber with an aperture on its front face. First, the transient responses between a source outside the cavity and an array of antennas within the cavity are measured. For wideband pulses, the signals spread over a time much longer than the initial pulse length because of the reverberation within the cavity. The signals are then flipped in time and re-emitted. Due to the reversibility of the wave in the propagation medium, the time-reversed field focuses both in time and space at the initial source position. The gain in amplitude of

Electromagnetic Time Reversal: Application to Electromagnetic Compatibility and Power Systems,
First Edition. Edited by Farhad Rachidi, Marcos Rubinstein and Mario Paolone.
© 2017 John Wiley & Sons, Ltd. Published 2017 by John Wiley & Sons, Ltd.
Companion Website: www.wiley.com/go/rachidi55

the focused signal is linked to the time compression of the transient response and can therefore be several orders higher than the amplitude generated using a beamforming method without the chamber. This method is demonstrated in the microwave range (1–4 GHz). We study the properties of the focal spot with respect to different experimental parameters such as the number of antennas, the aperture, and the size of the cavity or the source polarization. The one-bit time reversal method is also investigated to enhance the amplitude of the focused signal. Finally, we show an extension of the method to focus a pulse at any position outside the cavity from the knowledge of the transient responses on the aperture area.

4.2 Introduction

Focusing ultrawideband pulses with high resolution is commonly achieved with a beamforming method applied to an array of antennas. This method is used in a wide range of applications such as communications and radar imaging. It consists in applying a time delay to each antenna that is proportional to the distance between the antenna and the focal spot. Assuming free-space propagation, the emitted pulses interfere constructively at the focal point and the amplitude of the pulse at the focus is proportional to the number of antennas of the array. Obtaining high-amplitude pulses may therefore require a large number of emitting antennas.

Alternatively, time reversal (TR) is an adaptive focusing technique employed to focus waves in space and time [1]. TR takes advantage of the invariance property with time of the wave equation in a non-dissipative heterogeneous medium. First developed in the early 1990s, the methods based on TR are now widely used in acoustics [2–5], ultrasounds [6–9], electromagnetism [10–13], optics [14–16], and with water waves [17, 18]. A TR experiment consists in two steps. First, the temporal wavefield generated by a source is recorded simultaneously by an array of antennas, the so-called time-reversal mirror (TRM). The signals in a given time-window ΔT are then time-reversed, i.e., flipped in time, and re-emitted by the TRM. This naturally generates a wavefront converging on the initial source location.

Unlike beam-steering methods for which scattering within the medium decreases the focusing capabilities, TR takes advantage of the complexity of the medium to obtain a high-amplitude short pulse at the focus with a resolution which is improved in comparison to free space propagation. The contributions of the different optical paths within the medium indeed interfere constructively, both spatially and temporally. In a complex medium such as a multiple scattering medium or a reverberating cavity, the size of the focal spot is governed by the field correlation length rather than the aperture of the TRM [19–22]. Fine resolution can therefore be reached even with a small number of elements of the TRM. Focusing waves on a focal spot equal to the Rayleigh limit of $\lambda/2$ has been first demonstrated in a high Q-factor cavity with elastic waves and a single transducer [2]. Obtaining such a fine resolution requires broadband signals to be recorded on long time-windows. Long-travel paths corresponding to multiple scattering events are encapsulated in the *coda* of the recorded signals and reach the TRM at long delay times. In the second step of the TR experiment, those paths are re-excited and provide a high spatial diversity that leads to a small focal spot. In the time domain, the TR signal obtained at the focal spot is the autocorrelation of the transient response so that a pulse as short as the initial one is obtained at the focus.

The focusing performance of TR in complex media can be characterized by the ratio between the amplitude of the peak at the focus and the amplitude of the residual spatial and temporal sidelobes. This ratio has been shown to be equal to the number of spatiotemporal degrees of freedom N, which is the product of the number of spatial and temporal degrees of freedom, N_a and N_f, respectively. Assuming that the spacing between two antennas of the TRM is larger than half a wavelength, N_a reduces to the number of emitting antennas. The number of temporal degrees of freedom N_f is linked to the number of statistically independent frequency ranges within the bandwidth and can be expressed as $N_f = \Delta\omega/\delta\omega$. Here $\Delta\omega$ is the bandwidth of the signal and $\delta\omega$ is the typical width of the spectral field-correlation function. In a cavity, the envelope of the transient signal decays exponentially with time. When the decay time τ_d is much smaller than the Heisenberg time τ_h of the cavity (i.e., the average of the inverse of the modal density), $\delta\omega = \tau_d^{-1}$ and $N_f = \Delta\omega\tau_d$. However, when

$\tau_h \ll \tau_d$, the number of spectral degrees of freedom saturates and $N_f \sim \Delta\omega\tau_h$. A broadband field with $\Delta\omega \gg \delta\omega$ couples to a large number of modes of average linewidth $\Gamma = \tau_d^{-1}$ so that the intensity at the focus is greatly enhanced due to constructive interference of the modes. On the other hand, the background intensity results from the destructive interference of the modes. Those considerations highlight the influence of the bandwidth and shows that spatial and temporal information are inextricably linked in a TR experiment.

Using the time compression property of TR within a reverberation medium is therefore a potential solution to generate high-amplitude pulses with a small number of emitters. This was first demonstrated with ultrasounds by Montaldo *et al.* in 2001 [23]. The combination of TR and a few transducers fastened to a solid waveguide made of aluminum has led to shock wave lithotripsy with low-power electronics. A short pulse was emitted outside the waveguide and long signals were recorded on an array of transducers glued at the end of the waveguide. The time-reversed signals generated a high-amplitude acoustic pulse at the initial emission position. The authors also used the 1-bit TR method, which consists in setting the amplitude of the normalized signal to 1 or −1 depending on the sign of this one [24].

In electromagnetism, reverberation chambers (RCs) are commonly used for electromagnetic compatibility or antenna characterization. Thanks to multiple scattering on the RC walls, the field generated within the cavity is naturally diffuse. At frequencies much higher than the first resonance, the RC provides a statistically uniform field [26]. Here we use a leaky RC to obtain high amplitude pulses outside the RC [25]. A small aperture on the front face of the RC indeed makes it possible to focus electromagnetic waves at any desired position.

4.3 Measurements

4.3.1 Experimental Setup

The RC shown in Figure 4.1a is a parallelepiped cavity of dimensions $1.8 \times 1.2 \times 1.1$ m^3 with walls covered with aluminum. The size of the aperture on the front face of the RC can be

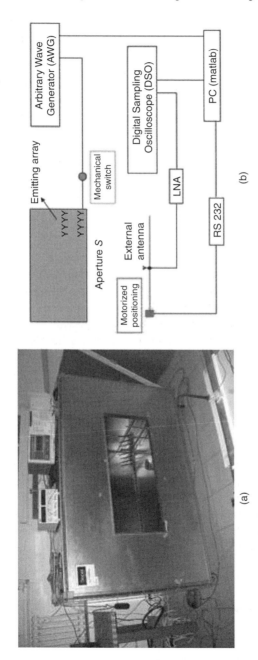

Figure 4.1 (a) Picture of the reverberation cavity. (b) Schematic view of the experimental setup.

modified by adding sheets of aluminum. Inside the chamber, the TR mirror (TRM) consists of eight half-wavelength omnidirectional antennas working in the frequency range 2.2–3.2 GHz. Those frequencies are well above the lowest usable frequency of the RC. The spacing between two antennas is equal to 10 cm. They are connected to an arbitrary wave generator (AWG) through an eight-channel electromechanical switch. The sampling rate of the AWG is 10 GS/s. Outside the RC, a motorized positioning antenna is located at 90 cm in front of the aperture. The signals are recorded using a digital sampling oscilloscope (DSO), with a sampling rate of 20 GS/s.

4.3.2 Time Compression

In a first step, a short signal with a carrier frequency of $\omega_0 = 2.7$ GHz modulated by a 1 ns Gaussian envelope is transmitted from an antenna inside the RC and the field is recorded with the external antenna outside the RC. The signal is normalized so that the maximum peak-to-peak output voltage is 1 V. Note that a basic TR experiment would involve generating the pulse with the external antenna and recording the field with the TRM. Nevertheless, due to the reciprocity theorem, the same signal is obtained by emitting the short pulse at the TRM and recording the field at the external antenna. This therefore makes it possible not to modify the connections of the AWG and the DSO between the two steps of the TR experiment.

The transient response $s(t)$ for a single emitting antenna is shown in Figure 4.2a for an aperture $S = 0.4$ m^2. Due to reverberation within the cavity, the signal spreads over more than 400 ns with an envelope that decays exponentially with time. This corresponds to a path length of more than 100 m before wave extinction. The transient signal is then flipped in time, normalized, and re-emitted. The TR field recorded on the external antenna is plotted in Figure 4.2b. A gain of 32 dB is obtained between the amplitude of the TR peak and the maximum of the transient response for an aperture of the cavity of $S \sim \lambda_0^2$.

4.3.3 Number of Antennas and Influence of the Bandwidth

We first study the dependence of the TR peak with respect to the number of antennas of the TRM. Due to the linearity of the TR

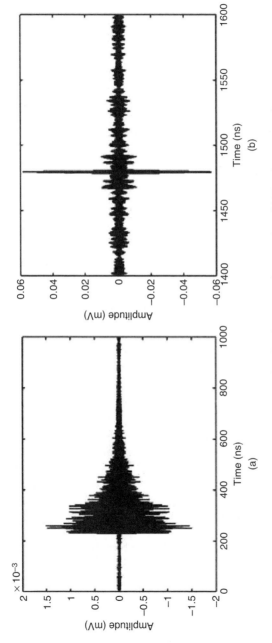

Figure 4.2 (a) Transient response of the reverberation chamber for a 1 ns pulse at 2.7 GHz; (b) TR pulse compression.

process, we find that the pulse amplitude increases linearly with the number of antennas (see Figure 4.3a). We have also studied the TR peak amplitude as a function of the duration of the initial pulse or, in other words, with respect to the bandwidth. In a TR experiment, all the frequencies within the bandwidth are added in phase at the time where the waves focus. Consequently, increasing the bandwidth enhances the amplitude of the focused pulse. A TR mirror is a generalization of the concept of the phase conjugation mirror working not only work at one frequency but also on an overall bandwidth [27]. In Figure 4.3b, TR signals are displayed for different initial pulse lengths. As expected, we observe that higher amplitudes are obtained for the shorter initial pulses. However, due to the finite bandwidth of the antennas, the TR pulse duration is limited to 1 ns. Consequently, the amplitude of the TR peak reaches a plateau (at -10 dBm) for very short initial pulses.

4.3.4 Size of the Focal Spot

The focal spot is measured thanks to the motorized positioning system linked to the external antenna. In Figure 4.4, the size of the lateral focal spot is plotted for two different sizes of the chamber apertures. The -3 dB focal spot widths are seen to be 150 mm and 400 mm for apertures with horizontal dimensions of 52 cm and 16 cm, respectively. In both cases, the vertical aperture is equal to 40 cm. These results are in agreement with the diffraction theory, which predicts that the focal spot size is roughly equal to $\lambda_0 R/L$, where L is the lateral size of the aperture and R is the distance between the external antenna and the aperture. The wavelength λ_0 is very close to the wavelength corresponding to the central frequency of the initial pulse.

4.3.5 One-Bit Time Reversal

The amplitude of the TR pulse can be enhanced by applying the one-bit TR method introduced by Derode *et al.* [24]. To this end, the amplitude of the re-emitted signal is set to 1 or -1 depending on whether the sign of the transient response is positive or negative. In comparison with a classical TR, Derode *et al.* showed that the spatiotemporal compression is barely modified but the

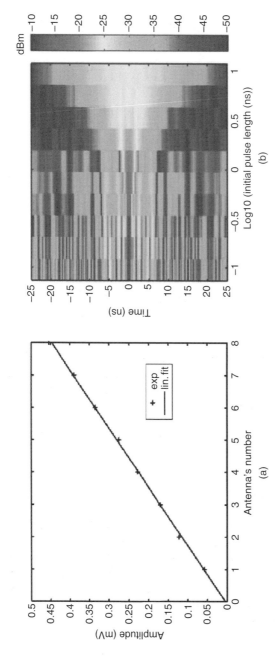

Figure 4.3 (a) Amplitude of the TR peak with respect to the number of antennas composing the TR mirror. (b) Signals after TR for different initial pulse lengths.

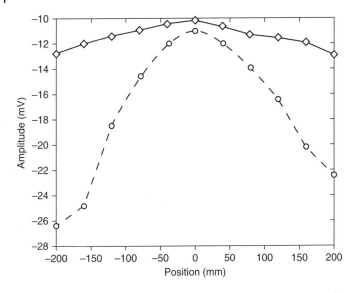

Figure 4.4 Focal spot for horizontal apertures of 52 cm (dashed line) and 16 cm (bold line).

amplitude of the pulse increases. The one-bit TR indeed mitigates half of the exponential decay of the transient signal by giving stronger weight to long delay times. We find that the one-bit TR pulse amplitude is enhanced by a factor ~4 in comparison to classical TR. Hence, for a surface $S \sim \lambda_0^2$, the gain between the TR peak and the maximum transient signal reaches 46 dB with a one-bit TR.

4.3.6 Aperture

We observe in Figure 4.4 that the smaller aperture gives the higher peak amplitude. This enhancement for small apertures is confirmed in Figure 4.5 in which the TR amplitude is shown as a function of the aperture S of the RC. This amplitude is plotted for values of S ranging from 0.02 m^2 to 0.4 m^2. A maximum is reached for a typical aperture $S \sim \lambda_0^2$. In the following section, we develop a model to study the pulse amplification and the influence of the aperture area of the RC.

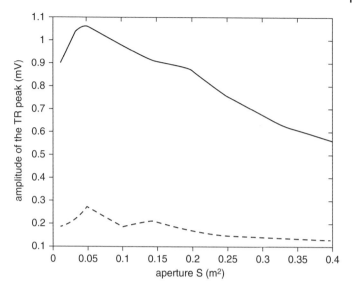

Figure 4.5 Amplitude of the one-bit (bold line) and "classical" (dashed line) TR focusing peak as a function of the aperture.

4.4 Theoretical Model

In the first step of the TR experiment, the signal transmitted through the aperture for a pulse emitted within the cavity for a given polarization i is measured. The electric field radiated through the aperture at a distance R is given by $E_{i,rad}(R, t) = h_{i,rad}(R, t) \otimes e(t)$. Here $h_{i,rad}(R, t)$ is the impulse response, i.e., the response to a Dirac excitation $\delta(t)$, and $e(t)$ is a short pulse of bandwidth $\Delta \omega$ centered on $\omega = \omega_0$. In the following $e(t)$ is normalized such as $e(t = 0) = e_{max} = 1$. This normalization is of importance to compare the amplitude of the pulse obtained with TR to the amplitude of the transient signal obtained in the first step. The transmitted signal $E_{i,rad}(R, t)$ is time-reversed, normalized by its maximum and re-emitted by the antenna inside the cavity. We thus obtain

$$E_{TR}(R, t) = \frac{h_{i,rad}(R, t) \otimes h_{i,rad}(R, -t) \otimes e(-t)}{\max[h_{i,rad}(R, t) \otimes e(t)]} \tag{4.1}$$

4.4.1 Average Amplitude of the Transient Signal

We first evaluate E_{max}, which is the average of the maximum of $E_{i,rad}(R, t)$. The impulse response radiated through the aperture can be derived from the Huygens–Fresnel principle. We note $h_i(r, t)$, the impulse response function of the RC evaluated at location r within the RC, and we first derive its autocorrelation. Using the Breit–Wigner formula, $h_i(r, t)$ can be written as the superposition of resonances with central frequencies ω_n, linewidth γ_n, and corresponding wavefunctions $\psi_{ni}(r)$:

$$h_i(r, t) = \sum_n \Psi_{ni}(r) \exp(i\omega_n t) e^{-\gamma_n t/2} \tag{4.2}$$

In the case of an open system such as the RC under consideration, the distribution of the linewidths γ_n is narrow so that $\gamma_n \sim \gamma = 1/\tau_d$ [28]. The decay time τ_d results from all the loss processes that are encountered in the RC. They not only include the flux radiated through the aperture but also the skin effect of the cavity boundaries, absorption by dielectric materials inside the cavity, and absorption of waves back-scattered to the emitting antenna. The coefficient $\Psi_n(r)$, which is the field amplitude of the nth mode of the cavity at location r, is a random variable with an autocorrelation function given by

$$\left\langle \Psi_{ni}(r) \Psi^*_{n'j}(r') \right\rangle = \frac{1}{3} \left\langle E_0^2 \right\rangle \operatorname{sinc}\left(\frac{\omega_n \|r - r'\|}{c_0} \right) \delta_{ij} \delta_{nn'}$$

Here c_0 is the speed of light. The factor $1/3$ takes into account the three polarizations that are statistically equivalent in the diffuse field model. The autocorrelation function of $h_i(r, t)$ is found from an integration over the bandwidth of the cardinal sine function and therefore reduces to a rectangular function:

$$\left\langle h_i(r, t) h_j^*(r, t) \right\rangle = \frac{\left\langle E_0^2 \right\rangle c_0 e^{-(t+t')/2}}{3\|r - r'\|} \Pi \left(\frac{c_0(t - t')}{\|r - r'\|} \right) \delta_{ij}. \tag{4.3}$$

For $r = r'$, Equation (4.3) gives

$$\left\langle h_i(r, t) h_j^*(r, t') \right\rangle = \left\langle E_0^2 \right\rangle e^{-(t+t')/2} \delta(t - t') \delta_{ij}/3$$

Assuming that the length of the pulse $e(t)$ is much shorter than the decay time τ_d, the root mean square (RMS) of the electric field within the cavity, $\langle E_i^2(t) \rangle = \langle [h_i(r,t) \otimes e(t)]^2 \rangle$, is estimated to be

$$\langle E_i^2(t) \rangle = \frac{\langle E_0^2 \rangle e^{-t/\tau_d}}{3} \int e^2(\tau) d\tau \tag{4.4}$$

The amplitude factor $\langle E_0^2 \rangle$ can be determined from the energy E at time $t = 0$ within the cavity. Because the field is assumed to be diffuse $E = V\epsilon \langle E_0^2 \rangle \int e^2(\tau) d\tau$, where V is the volume of the cavity and ϵ is the permittivity of air [29]. This energy is equal to the energy injected by the feeding antenna excited by a short pulse $e(t)$ that is maximum at $t = 0$, i.e., $P_{max} \int e^2(\tau) d\tau / e(0)^2$, where P_{max} is the maximum instantaneous power injected to the cavity by the RF generator. Since $e(t)$ is normalized, $\langle E_0^2 \rangle = P_{max}/[V\epsilon]$.

We then estimate the RMS of the field radiated through the aperture outside the cavity. We assume that (i) the length of the pulse $e(t)$ is large compared to its central period but much smaller than τ_d and (ii) the focal point is located in the far-field of the aperture at a distance R. Using the Huygens–Fresnel principle, straightforward calculations lead to

$$\langle E_{i,rad}^2(t) \rangle = \frac{2 \langle E_0^2 \rangle}{3} \frac{k_0^2 [\int e^2(\tau) d\tau]}{(4\pi R)^2}$$
$$e^{-t/\tau_d} \iint \mathrm{sinc}(k_0 ||r - r'||) d^2r \, d^2r' \tag{4.5}$$

In this expression, $k_0 = 2\pi/\lambda_0$. The double integral is performed over the cavity aperture area. $\langle |E_{i,rad}(t)|^2 \rangle$ is seen to be related to the autocorrelation of the field on the aperture S. In the following, we note $S_{eff} = [\iint \mathrm{sinc}(k_0 ||r - r'||) d^2r \, d^2r']/S$ and S_{eff} has the dimensions of a surface. We then obtain

$$E_{max}^2 = \frac{2 \langle E_0^2 \rangle}{3} \frac{k_0^2 \left[\int e^2(\tau) d\tau \right]}{(4\pi R)^2} S_{eff} S \tag{4.6}$$

4.4.2 Gain Obtained Using TR

The field $E_{i,rad}(R,t)$ is time-reversed, normalized with respect to its maximum and re-emitted by the antenna inside the cavity using the full instantaneous emitter power capacity. Here we take benefit of the reciprocity theorem to avoid the exchange between the emitter and receiver positions. From the Huygens–Fresnel principle and Equation (4.1), it can be shown that

$$E_{TR}(t) = \frac{2\langle E_0^2\rangle}{3E_{max}} \frac{k_0^2 e(-t)\left[\int e^{-\tau/\tau_d}d\tau\right]}{(4\pi R)^2} S_{eff} S \tag{4.7}$$

We first evaluate the ratio between the amplitude of the time-reversed pulse, $E_{TR} = E_{TR}(t = 0)$, and the maximum of the transient field E_{max}:

$$\frac{E_{TR}}{E_{max}} = \frac{\tau_d}{\int e^2(\tau)d\tau} \sim \tau_d \Delta\omega \tag{4.8}$$

We observe in Equation (4.8) that the enhancement of the pulse is given as the product of the decay time of the cavity time and the pulse bandwidth. This result is in agreement with the discussion in the introduction about the number of degrees of freedom.

The instantaneous gain can also be derived from Equation (4.7):

$$G = \frac{4\pi R^2 E_{RT}^2}{\eta P_{max}} = G_a G_c \tag{4.9}$$

where

$$G_a = \frac{4\pi S_{eff}}{\lambda_0^2} \quad \text{and} \quad G_c = \frac{c_0 S \tau_d^2 \Delta\omega}{6V} \tag{4.10}$$

In this equation G_a can be interpreted as the gain of an equivalent parabolic antenna of size S_{eff}. G_c is the gain due to pulse compression in the TR process. Assuming that the main contribution to τ_d is the flux radiated through aperture gives $\tau_d^{-1} \sim c_0 S/V$ (Sabine's formula), Equation (4.10) leads to

$$G = \frac{2\pi V \Delta\omega}{3\lambda_0^2 c_0} \frac{S_{eff}}{S} \tag{4.11}$$

Equation (4.11) shows that the gain depends on the ratio between the effective surface and the aperture of the cavity. The ratio S_{eff}/S decreases from unity for an aperture typically small compared to λ_0^2 since $S_{eff}/S \sim 0.6\lambda_0^2/S$ for $S \gg \lambda_0^2$. A small aperture hence leads to a high TR amplitude. This is in agreement with measurements of the TR amplitude, E_{TR}, as a function of S for large S (see Figure 4.5) since E_{TR} decreases with S. Nevertheless, a maximum in measurements is reached for an aperture $S \sim \lambda_0^2$. For subwavelength apertures, the decay time τ_d is indeed dominated by other processes that do not contribute to the increase of the TR amplitude, such as the leakage through the walls of the cavity. As a result an optimum aperture is found in the measurements.

4.5 Comparison with a Directive Antenna

4.5.1 Comparison with a Directive Antenna

To evaluate the performances of our system, we compare the TR peak with the amplitude of the pulse generated by a directive antenna with the same aperture. To this end, we replace our time-reversal mirror made of dipoles by a single horn antenna of aperture 0.4 m^2, working between 1 and 18 GHz. The width of the beam is equal to 35.94° at 3 GHz. The gain of our system is determined from the two following steps. First, the horn antenna, in the absence of the RC, is pointed towards the focal spot at a distance R. Second, the horn antenna is located within the RC and the focal spot is measured at the same distance R from the aperture of the RC. In both cases, TR experiments are performed so that the influence of scattering events outside the RC is minimized. We obtain a gain of 18 dB on the amplitude of the TR peak at a distance $R = 2.50$ m.

4.5.2 Autosteering Properties of the System

In addition to the pulse compression effect which strongly enhances the focused signal, another advantage of the system is the electronic steering capability. A directive antenna needs to be pointed mechanically towards the focal spot whereas our system focuses on the initial source position even though its position is off-centered. Obviously the gain between the two methods is all the more important that the source is off-centered. This is illustrated in Figure 4.6a. For the single horn in the absence of the RC, the amplitude decreases rapidly as a function of the angle θ between the antenna and the direction of the focal spot. On the other hand, E_{TR} with the RC is almost constant until $\theta = 40°$.

The system is furthermore autoadaptative in polarization. Due to multiple scattering within the RC, the three polarizations are statistically equivalent. Hence, a TR pulse with arbitrary polarization can be obtained independently of the polarization of the antenna inside the chamber. This is demonstrated in Figure 4.6b. E_{TR} is seen to be the same for parallel polarizations and orthogonal polarizations. In contrast, a decrease of ~10 dB is seen in the absence of the RC.

4.6 Discussion

4.6.1 Focusing at Any Location

In this study, TR focusing requires recording first the transient response between the TRM and the focal point. This is achieved by emitting a short pulse with an external antenna, but this method does not make it possible to focus radiations at any location or on a passive target. This issue may be overcome using the synthetic TR method, which has first been demonstrated for focusing in acoustics using a single transducer coupled to a solid cavity immersed in water [30]. The authors have shown that the transient response between the transducer and any point in space can be known when a library of a few reference impulse responses is obtained by scanning the field on the front face of the cavity. The time-reversed signal that will converge towards any desired location can then be constructed synthetically with a beamforming method.

The same idea has been applied in electromagnetism by Hong *et al.* with a parallelepiped RC [31]. They first modified the shape of the aperture of the RC. Instead of a single rectangular aperture, they used discretized small apertures distributed over the front face of the cavity (see Figure 4.7). In a first step, the field emitted by the cavity feed is scanned on those apertures.

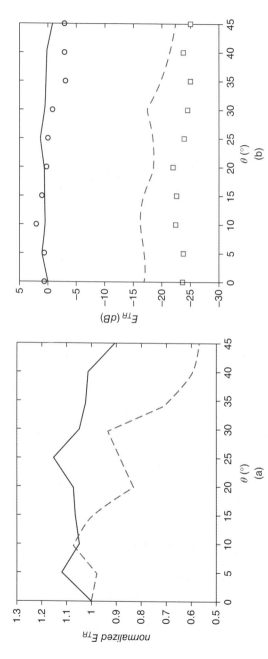

Figure 4.6 (a) The amplitude E_{TR} of the amplitude using TR with the RC (solid line) and in the absence of the RC (dashed line) is plotted with angle θ between the aperture and the direction of the focal spot. Both curves are normalized for $\theta = 0°$. (b) E_{TR} in dB, with the RC and in the absence of the RC with θ for the same polarization (solid and dashed lines, respectively) and orthogonal polarization (dots and squares, respectively).

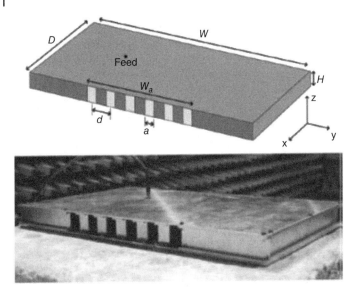

Figure 4.7 Radiated structure constructed for impulse response simulation (top) and experiment (bottom). Adapted from [31].

Focusing at any location in space can then be obtained by applying appropriate time delays to the time reversal of those signals. The time delays are calculated from the distance between those apertures and the desired focal spot. They have shown in measurements that this method makes it possible to accurately point the beam at any desired angle over a wide range of scanned angles. In comparison with a single aperture in the middle of the sample, this geometry also enhances the resolution of the system since the aperture of the equivalent antenna is increased. Nevertheless, those distributed apertures may lower the Q-factor of the cavity and hence lower the gain.

4.6.2 Obtaining High Resolution and High Q-Factor

A promising design to enhance the intrinsic performances of the system involves using a subwavelength diffraction grating instead of a regular aperture. Dupré *et al.* have shown that in such a configuration the modes of the cavity are perfectly transmitted to the exterior, even for high Q-factor cavities [32]. The

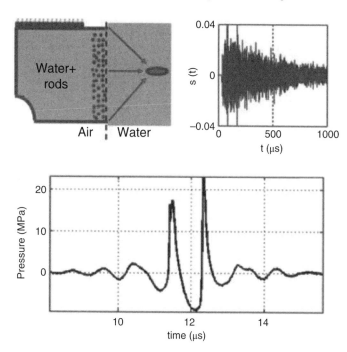

Figure 4.8 (top) From left to right: concept scheme of the tunable TR cavity and a typical signal obtained by an FDTD simulation. (bottom) Shock wave obtained with the TR cavity at maximum output voltage. Adapted from [9].

Q-factor can be tuned by changing the filling factor of the grating. This could have a great potential to get high-amplitude pulses with TR.

Another way to get both a high Q-factor and a large aperture is to embed a multiple scattering medium within the cavity. This configuration has been used in acoustics by Arnal *et al.* (see Figure 4.8) [9]. This design improves the quality factor of the cavity and at the same time provides a good transmission coefficient and a high aperture. The waves are therefore focused outside the cavity with fine resolution. The density of scatterers of the medium is chosen as a trade-off between the reverberation time of the cavity and the losses. In comparison with a conventional focusing method with the same probe, the authors have obtained a gain as high as 25 dB in amplitude and managed to generate

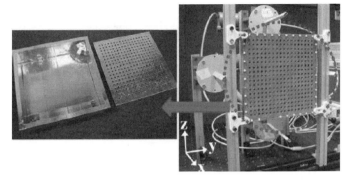

Figure 4.9 Overall view of the cavity imaging system including the transmitting cavity (a detailed view of the mode-mixing cavity structure can be seen on the left) and the four receiving probes. From [33].

a shock-wave with their low-voltage electronics. This may find applications in lithotripsy or histotripsy.

4.6.3 Radar Imaging

We finally would like to discuss the use of a similar cavity opened on its front face for radar imaging. Obtaining high-resolution images with only a few antennas has been demonstrated by Fromenteze *et al.* [33]. They have presented a three-dimensional imaging device based on a mode mixing cavity (see Figure 4.9). An emitting antenna is located inside a high Q-factor chaotic RC with periodic holes perforated on its front side. The radiated fields are scattered by the scene and measured with four receiving probes positioned around the cavity. The signals are measured on a wide frequency range over which the modes of the cavity and consequently the radiation patterns at the scene drastically change. Those patterns are known thanks to a calibration stage in which the near-field of the cavity is scanned over its aperture. The three-dimensional reflectivity map of the scene is then constructed using computational methods. Similarly to a TR focusing device, the resolution of the image is given by the diffraction limit associated with the cavity aperture instead of the antenna array aperture. The imaging system takes advantage of the spectral degrees of freedom provided by the chaotic cavity since the scene information is encoded on to its independent modes. The method has the potential to strongly accelerate the imaging process in comparison with synthetic aperture radar methods since it does not require moving the antennas and only a small number of antennas are used to compute high-resolution images.

4.7 Conclusion

In conclusion, the coupling of TR with a leaky cavity makes it possible to generate high-amplitude wideband signals with a single feed or a small number of antennas. The temporal compression of an impulse response spreading over a long time creates a pulse with enhanced amplitude. Using a diffuse field model, we have demonstrated that the maximum amplitude is obtained for cavities with a small aperture and therefore a large Q-factor. We have discussed various improvements of the method to achieve simultaneously a high Q-factor and a fine resolution. In addition to security applications, this method could be applied to a range of biomedical applications such as microwave hyperthermia for cancer treatment or breast cancer detection.

References

1 M. Fink, *Physics Today*, vol. 50, p. 34, 1997.
2 C. Draeger and M. Fink, *Physical Review Letters*, vol. 79, p. 407, 1997.
3 M. Fink and C. Prada, *Inverse Problems*, vol. 17, p. R1, 2001.
4 G. F. Edelmann, H. C. Song, S. Kim, W. S. Hodgkiss, W. A. Kuperman, and T. Akal, *IEEE Journal of Oceanic Engineering*, vol. 30, p. 852, 2005.
5 F. Lemoult, M. Fink, and G. Lerosey, *Physical Review Letters*, vol. 107, p. 064301, 2011.
6 A. Derode, A. Tourin, and M. Fink, *Physical Review E*, vol. 64, p. 036606, 2001.
7 M. Pernot, G. Montaldo, M. Tanter, and M. Fink, *Applied Physics Letters*, vol. 88, p. 034102, 2006.
8 O. Couture, J.-F. Aubry, M. Tanter, and M. Fink, *Applied Physics Letters*, vol. 94, p. 173901, 2009.

9 B. Arnal, M. Pernot, M. Fink, and M. Tanter, *Applied Physics Letters*, vol. 101, p. 064104, 2012.

10 G. Lerosey, J. De Rosny, A. Tourin, A. Derode, and M. Fink, *Applied Physics Letters*, vol. 88, p. 154101, 2006.

11 G. Lerosey, J. de Rosny, A. Tourin, and M. Fink, *Science*, vol. 315 p. 1120, 2007.

12 F. Lemoult, G. Lerosey, J. de Rosny, and M. Fink, *Physical Review Letters*, vol. 104, p. 203901, 2010.

13 M. Frazier, B. Taddese, T. Antonsen, and S. M. Anlage, *Physical Review Letters*, vol. 110, p. 063902, 2013.

14 D. Miller, *Optics Letters*, vol. 5, p. 300, 1980.

15 M. F. Yanik and S. Fan, *Physical Review Letters*, vol. 93, p. 173903, 2004.

16 S. Longhi, *Physical Review E*, vol. 75, p. 026606, 2007.

17 A. Przadka, S. Feat, P. Petitjeans, V. Pagneux, A. Maurel, and M. Fink, *Physical Review Letters*, vol. 109, p. 064501, 2012.

18 A. Chabchoub and M. Fink, *Physical Review Letters*, vol. 112, p. 124101, 2014.

19 R. Mallart and M. Fink, *Journal of the Acoustical Society of America*, vol. 96, p. 3721, 1994.

20 A. Derode, P. Roux, and M. Fink, *Physical Review Letters*, vol. 75, p. 4206, 1995.

21 P. Blomgren, G. Papanicolaou, and H. Zhao, *Journal of the Acoustical Society of America*, vol. 111, p. 230, 2002.

22 G. Bal and L. Ryzhik, *SIAM Journal on Applied Mathematics*, vol. 63, p. 1475, 2003.

23 G. Montaldo, P. Roux, A. Derode, C. Negreira, and M. Fink, *Journal of the Acoustical Society of America*, vol. 110, p. 2849, 2001.

24 A. Derode, A. Tourin, and M. Fink, *Journal of Applied Physics*, vol. 85, p. 6343, 1999.

25 M. Davy, J. de Rosny, J.-C. Joly, and M. Fink, *Comptes Rendus Physique*, vol. 11, p. 37, 2010.

26 D. A. Hill, *IEEE Transactions on Electromagnetic Compatibility*, vol. 40, p. 209, 1998.

27 A. Derode, A. Tourin, and M. Fink, *Ultrasonics*, vol. 40, p. 275, 2002.

28 U. Kuhl, H. Stöckmann, and R. Weaver, *Journal of Physics A: Mathematical and General*, vol. 38, p. 10433, 2005.

29 D. Hill, M. T. Ma, A. R. Ondrejka, B. F. Riddle, M. L. Crawford, and R. T. Johnk, *IEEE Transactions on Electromagnetic Compatibility*, vol. 36, p. 169, 1994.

30 N. Quieffin, S. Catheline, R. K. Ing, and M. Fink, *Journal of the Acoustical Society of America*, vol. 115, p. 1955, 2004.

31 S. K. Hong, V. Mendez, W. S. Wall, and R. Liao, *Antennas and Wireless Propagation Letters, IEEE*, vol. 13, p. 794, 2014.

32 M. Dupré, M. Fink, and G. Lerosey, *Physical Review Letters*, vol. 112, p. 043902, 2014.

33 T. Fromenteze, O. Yurduseven, M. F. Imani, J. Gollub, C. Decroze, D. Carsenat, and D. R. Smith, *Applied Physics Letters*, vol. 106, p. 194104, 2015.

5

Application of Time Reversal to Power Line Communications for the Mitigation of Electromagnetic Radiation

P. Pagani,[1] M. Ney,[1] and A. Zeddam[2]

[1] *Telecom Bretagne, Lab-STICC UMR CNRS 6285, Brest, France*
[2] *Orange Labs, Lannion, France*

5.1 Introduction

Power line communications (PLC) is a technology that uses the conventional electrical network as a communication medium [1]. This allows for the deployment of a quasi-ubiquitous communication network without the need for installing an additional infrastructure. The PLC technology is typically used in the indoor environment over the low-voltage (LV) infrastructure. In this environment, the attenuation between different outlets in a given electrical network is relatively limited, which allows for throughputs in the order of tens of Mbps. Such broadband (BB) PLC systems operate in a frequency band ranging from 2 MHz to 100 MHz depending on the standard. Among the most recent BB PLC specifications, one can cite IEEE 1901 [2], ITU-T G.9960 [3], and HomePlug AV2 [4]. Another application of PLC is to transmit command and control information over the LV or medium-voltage (MV) networks, using frequencies below 500 kHz. These systems fall within the class of narrowband (NB) PLC and allow

Electromagnetic Time Reversal: Application to Electromagnetic Compatibility and Power Systems,
First Edition. Edited by Farhad Rachidi, Marcos Rubinstein and Mario Paolone.
© 2017 John Wiley & Sons, Ltd. Published 2017 by John Wiley & Sons, Ltd.
Companion Website: www.wiley.com/go/rachidi55

for long-range communication in the Smart Grid environment [5]. The ITU-T G.9955 standard is an example of such NB PLC systems [6].

The PLC technology uses a transmission medium consisting of copper wires, connectors, and electrical equipment: outlets, switches, circuit breakers, etc. When the frequency of the transmitted signal increases up to the high-frequency (HF) band, these elements lead to a strong attenuation. In addition, the topology of the electrical network generates multiple propagation paths, which increases the frequency selectivity of the channel. The impedance of the loads connected to the network is also highly variable, depending not only on the location but also on the conductor (live or neutral) used for the measurement. Finally, the different conductors forming the electrical cable are not necessarily of the same length, in particular due to the presence of single-phase switches. Hence, in the HF band, the electrical network can be seen as an unbalanced transmission line. As a consequence, part of the differential PLC signal injected in the electrical network is converted into common mode current along the network. In turn, this common mode current generates non-negligible electromagnetic radiation [7]. This mechanism potentially yields to specific electromagnetic compatibility (EMC) issues, with respect to other services using the same frequency band. In the HF band, one can cite amateur radio (HAM) communications and short-wave (SW) radio broadcasting services.

This chapter focuses on the issue of EMC for PLC systems. This topic has been addressed, for instance, within the research project ICT FP7 OMEGA [8] and at a standardization level in the framework of the ETSI Specialist Task Force 410 [9]. The first countermeasure defined to protect users of the electromagnetic spectrum consists of imposing a strict limit to the authorized transmission power. Such emission masks are defined in regulation standards such as the Federal Communications Commission (FCC) Part 15 in the USA [10] and the standard EN 50561-1 in Europe [11]. In addition to spectrum regulation, different methods can be applied in order to minimize the undesired radiation of PLC systems. For instance, a complementary signal can be injected into the electrical network, using a particular waveform in order to cancel the perceived electrical field. This technique was tested in [12] and allows an effective reduction of the radiated power to protect a given potential victim. However, this technique can be optimized for a single location only and cannot be applied simultaneously for the entire environment. Another method presented in [13] consists in connecting additional plugs to the network outlets in order to reduce the impedance mismatch at the network extremities. The drawback of this technique is that is requires the deployment of additional hardware.

In this chapter, we investigate the signal processing technique of time reversal (TR), initially developed in the acoustic domain and then further extended to the field of electromagnetic propagation [14]. This technique consists of using the knowledge of the transmission channel characteristics in order to focus the received signal both in time and space. In the field of wireless communication, experiments using ultrawideband (UWB) signals showed that the obtained focusing effect allowed a reduction of the interference between radio systems [15, 16]. For PLC systems with a limited transmission power, time-reversal signal processing is a promising candidate to solve EMC problems. On the one hand, focusing the transmitted energy at the intended receiver location provides a channel gain, which can in turn be used to reduce the level of transmitted power. On the other hand, the reduction of the perceived signal level in other parts of the network, and especially along the copper wires, is expected to reduce the level of common mode current and hence to mitigate unintentional radiations.

The potential of TR to handle EMC issues for PLC systems is studied in this chapter. We detail the principles of TR and its application to wireline communications in Section 5.2. Section 5.3 describes a measurement experiment conducted in the 2–28 MHz frequency band, which corresponds to in-home BB PLC systems. Finally, the performance of the TR technique for the mitigation of the electromagnetic interference (EMI) is statistically assessed in Section 5.4.

5.2 Adaptation of Time Reversal to Power Line Communications

The development of the TR technique first appeared in the field of acoustic wave transmission [17, 18] and was later extended to the field of radio transmission [14]. The propagation of electromagnetic waves leads to the formation of multiple echoes, and such a multipath channel is particularly suited for exploiting the TR technique.

The TR technique consists of pre-filtering the transmitted signal using a filter adapted to the transmission channel. In the following, one assumes that the transmitter is placed at the space origin, and the location of the receiver is noted r_0. Denoting $H(f, r_0)$ as the channel transfer function (CTF) for any frequency f, the input filter $G(f, r_0)$ is proportional to $H(f, r_0)^*$, where $*$ represents the complex conjugate operator. More precisely, for an antenna placed at an arbitrary location r, the perceived CTF after applying the TR technique, $H_{RT}(f, r)$, is given by

$$H_{TR}(f, r) = G(f, r_0) \times H(f, r) = \frac{H^*(f, r_0)}{\sqrt{\int |H(f, r_0)|^2 df}} \times H(f, r)$$

(5.1)

One can note that the matched filter is normalized to a unit power. Several studies demonstrated that this technique simultaneously allows the transmitted signal to be focused around the intended receiver location, and reducing the received power at any other location in the environment [15, 16, 19]. In particular, applying Equation (5.1) to the receiver situated at location r_0 leads to

$$H_{TR}(f, r_0) = \frac{1}{\sqrt{\int |H(f, r_0)|^2 df}} \times |H(f, r_0)|^2$$

(5.2)

The perceived CTF after applying TR is hence proportional to the square of the amplitude of the effective CTF. This allows a gain in the total received power since the frequency selective channel is more optimally exploited.

When considering TR in the time domain, the propagation channel is represented by the channel impulse response (CIR), $h(\tau, r)$, for which Equation (5.1) translates to

$$h_{TR}(\tau, r) = \frac{h(-\tau, r_0)^*}{\sqrt{\int |h(\tau, r_0)|^2 d\tau}} \otimes h(\tau, r)$$

(5.3)

At the intended Rx location r_0, Equation (5.3) leads to

$$h_{TR}(\tau, r_0) = \frac{1}{\sqrt{\int |h(\tau, r_0)|^2 d\tau}} \times R_h(\tau, r_0) \qquad (5.4)$$

where $R_h(\tau, r_0)$ corresponds to the autocorrelation of the CIR $h(\tau, r_0)$. One can observe that applying TR corresponds to modifying the perceived CIR into the autocorrelation of the actual channel. When the environment provides rich scattering and multiple propagation paths, the CIR autocorrelation is generally composed of a strong peak at zero offset with low-level side echoes. As a consequence, the received energy is more focused in time. Reference [19] reports gains in the order of 5 dB when TR is applied to UWB channels.

In addition, for any location r different from r_0, applying TR generates a mismatch of the shaping filter with respect to the actual channel. The perceived CTF $H_{RT}(f, r)$ corresponds to the product of two independent CTFs, $H(f, r_0)^*$ and $H(f, r)$, each of them exhibiting an independent fading structure in the frequency domain. As a consequence, for receiving locations r different from the intended location r_0, the total received power is reduced. This effect is called spatial focusing in the area of wireless transmission. The spatial focusing parameter η is a figure of merit used to evaluate this phenomenon. It is given as a function of the distance $\|r - r_0\|$ as:

$$\eta = 10 \log_{10}(\max(|h_{TR}(\tau, r)|^2)) - 10 \log_{10}(\max(|h_{TR}(\tau, r_0)|^2)) \qquad (5.5)$$

References [19] and [20] reported on experimental measurements of wireless channels, where spatial focusing factors in the order of -10 dB were observed.

The analysis of the TR technique in the wireless field sheds light on two features that are also desirable for powerline communications. First, the total received power can be increased at the intended Rx location. Without modifying the Tx power level, this would lead to an enhanced performance of a TR PLC system. In our context of radiation mitigation, the TR channel gain can also be converted into a reduction of the Tx power level, while maintaining the same communication performance. Second, the

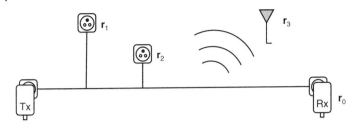

Figure 5.1 Application of the TR technique to powerline transmission.

application of TR reduces the received power at any location except at the intended Rx. Both of the aforementioned features are expected to reduce the level of undesired radiated energy in a PLC system.

More specifically, one considers an indoor powerline network where a Tx PLC modem is placed at the origin and a Rx PLC modem is placed at a location r_0. The open literature on PLC channel characterization reports that this environment produces a number of echoes due to the network topology and the varying impedances at its different nodes [21–24]. As a result, the structure of the PLC channel is very similar to a radio channel, which provides some confidence for a successful application of TR to PLC.

With reference to Figure 5.1, by filtering the transmitted signal using the knowledge of the channel state information (CSI), one can expect a gain in the received power at the outlet situated at location r_0. Simultaneously, the signal received at any other outlet, for instance at locations r_1 and r_2, should be attenuated. More interestingly, this matched filtering will also affect the signal radiated from the electrical cables and received at an arbitrary location in the surrounding environment, such as r_3 for instance.

The application of the TR technique to wireline communication was first presented in [25]. Promising results were obtained from a limited set of measurements performed in a laboratory environment using a dedicated time-domain sounding equipment. In the following, we will assess the performance of TR applied to PLC using a new set of experimental data collected in the frequency domain in a more realistic in-home environment.

5.3   Experimental Study of Radiation Mitigation

5.3.1   Measurement Campaign

In order to statistically evaluate the performance of the TR technique in terms of mitigation of the EMI in a wireline system, both CTF and radiation measurements were carried out in three individual houses in France. Figure 5.2 presents the experimental setup used for this campaign. The collected data corresponds to measurements of the channel S21 parameter, using a vector network analyzer (VNA) Agilent E5071C. The passive PLC couplers were initially realized within the ETSI Specialist Task Force 410 [26]. The Tx coupler was connected to an outlet at the origin and the Rx coupler was connected to another outlet of the electrical network at location r_0. We first measured the CTF $H(f, r_0)$ corresponding to the wireline Tx–Rx propagation channel. As a second step, the CTF $H(f, r_3)$ was measured between the Tx coupler and a receiving antenna situated at an arbitrary location in the vicinity of the electrical network. We used a biconical antenna Schwarzbeck EFS 9218. The small size of this active antenna allowed a convenient displacement for making measurements at different places in the considered environments. Figure 5.3 presents a picture of the experimentation, where both the injection coupler and the receiving antenna are visible.

The considered frequency band extends from 2 MHz to 28 MHz, which corresponds to the band used by most of the

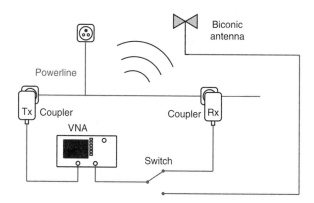

Figure 5.2 Equipment used in the experimental setup.

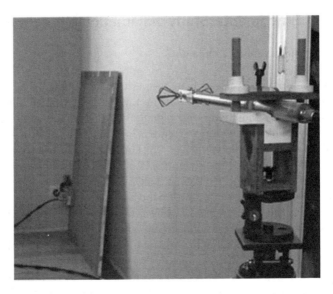

Figure 5.3 Picture of the experimentation.

current commercial PLC modems. It should be noted that the most recent PLC specifications, such as Home Plug AV2, make use of the protective earth (PE) conductor in addition to the classical live (L) and neutral (N) wires of the electrical network. Therefore, in our experiment, a signal was successively injected onto the three possible differential injection ports: L–N, L–PE, and N–PE. In total, 54 couples of CTFs $H(f, r_0)$ and $H(f, r_3)$ were used as a basis for our statistical analysis.

It can be noted that our experimental setup was developed for BB PLC systems for practical reasons. Similar investigations could be conducted for other wired communication media, such as the outdoor electrical network used by the NB PLC technology or the telephone copper wires used for digital subscriber line (DSL) access.

5.3.2 Data Processing

In order to analyze the impact of applying TR to PLC communications, the collected experimental data were post-processed to extract several parameters, as introduced in [25] and described in the following paragraphs.

From a single measurement of the PLC CTF $H(f, r_0)$, the mean channel attenuation before applying TR is defined in dB as

$$\overline{H(r_0)} = 10 \log_{10} \left(\frac{1}{f_{max} - f_{min}} \int_{f_{min}}^{f_{max}} |H(f, r_0)|^2 \, df \right) \quad (5.6)$$

where $f_{min} = 2$ MHz and $f_{max} = 28$ MHz represent the minimum and maximum frequencies, respectively, considered in our study.

The mean channel attenuation $\overline{H(r_0)}$ can be regarded as the attenuation perceived by a receiver capable of extracting all the available power in the system frequency band. This is the case for most of the commercial PLC systems using orthogonal frequency division multiplexing, for instance HomePlug AV devices.

Similarly, using the CTF that the receiver perceives after applying TR filtering, as defined in Equation (5.2), one can compute the mean channel attenuation after applying TR:

$$\overline{H_{TR}(r_0)} = 10 \log_{10} \left(\frac{1}{f_{max} - f_{min}} \int_{f_{min}}^{f_{max}} |H_{TR}(f, r_0)|^2 df \right) \quad (5.7)$$

From Equations (5.6) and (5.7), it is possible to define for each measurement the TR channel gain G_{TR}, which corresponds to the gain in total received power due to the application of the TR filter. G_{TR} is given in dB as

$$G_{TR} = \overline{H_{TR}(r_0)} - \overline{H(r_0)} \quad (5.8)$$

In order to study the effect of TR on the undesired EMI, we assume that a PLC modem injects a signal with a power spectral density (PSD) $P_{in} = -55$ dBm/Hz on the electrical network. This level corresponds to the feeding level currently authorized by the international regulation bodies. It is then possible to compute the electrical field power density at location r_3, $E(f, r_3)$, in dBμV/m, from the CTF $H(f, r_3)$ measured between the Tx coupler and the receiving antenna [26]:

$$E(f, r_3) = P_{in} + 20 \log_{10}(|H(f, r_3)|) + 107 + AF(f) \quad (5.9)$$

where 107 corresponds to the conversion from dBm to dBμV and $AF(f)$ represents the antenna factor.

Finally, the mean radiated power density over the considered frequency band, $\overline{S(r_3)}$, is given in dB (W/m^2) as

$$\overline{S(r_3)} = 10 \log_{10} \left(\frac{1}{f_{max} - f_{min}} \int_{f_{min}}^{f_{max}} \frac{1}{120\pi} |E(f, r_3)|^2 df \right) \quad (5.10)$$

where $E(f, r_3)$ is expressed in V/m and the term 120π corresponds to the free-space impedance in Ω.

The quantities $E(f, r_3)$ and $\overline{S(r_3)}$ can also be computed in the case where TR filtering is applied. For this, Equations (5.9) and (5.10) are transformed by replacing the CTF $H(f, r_3)$ by the CTF $H_{TR}(f, r_3)$

$$E_{TR}(f, r_3) = P_{in} + 20 \log_{10}(|H_{TR}(f, r_3)|) + 107 + AF(f) \quad (5.11)$$

and

$$\overline{S_{TR}(r_3)} = 10 \log_{10} \left(\frac{1}{f_{max} - f_{min}} \int_{f_{min}}^{f_{max}} \frac{1}{120\pi} |E_{TR}(f, r_3)|^2 df \right)$$
$$(5.12)$$

Finally, it is possible to compute an EMI reduction coefficient R_{TR} in dB that determines how much the mean radiated power density is reduced due to the application of the TR filter:

$$R_{TR} = \overline{S(r_3)} - \overline{S_{TR}(r_3)} \quad (5.13)$$

5.4 Results and Statistical Analysis

5.4.1 Measurement Example

Figure 5.4 provides an example of CTF $H(f, r_0)$ measured in a particular configuration of Tx and Rx modems. The measured CTF before applying TR filtering is represented by the solid curve. One can observe both a decreasing magnitude of the received signal with increasing frequency and a strong variation of the signal for adjacent frequencies. This latter phenomenon

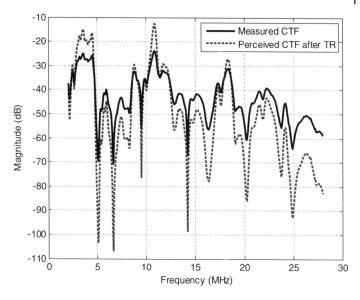

Figure 5.4 Channel attenuation before and after TR processing.

corresponds to the frequency fading due to the multiple propagation paths between the Tx and Rx couplers. In this example, the mean channel attenuation is 35.1 dB.

The dotted curve corresponds to the perceived CTF $H_{TR}(f, r_0)$ after applying TR. One observes that the least attenuated frequencies benefit from a power increase, while the frequency notches tend to deepen. In particular, for all frequencies where the CTF magnitude is larger than the mean value $\overline{H(r_0)}$, the perceived channel is improved by the application of TR. More precisely, the increased received power is observable, for instance, in the frequency bands from 2.8 MHz to 4.5 MHz and from 10.2 MHz to 11.4 MHz. When computing the mean channel attenuation, the application of TR leads to a value of 27.5 dB. Therefore, the gain in total received power linked to the TR filtering process is $G_{TR} = 7.6$ dB.

Figure 5.5 represents the electrical field measured by the receiving antenna. Without applying TR, the mean radiated power density can be computed as $\overline{S(r_3)} = -60.5$ dB (W/m^2). In this particular example, the effect of the application of the TR

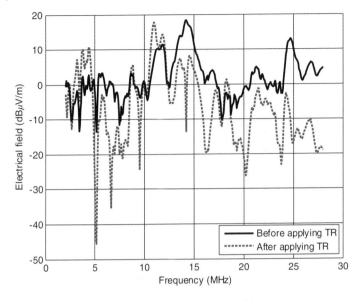

Figure 5.5 Electrical field before and after TR processing.

filter is particularly beneficial for frequencies above 12.5 MHz, where the electrical field is reduced for the large majority of the frequency spectrum. Even if some radiation increase is observable in the lower part of the spectrum, this effect is compensated, on average, as the mean radiated power density after TR is given as $\overline{S_{TR}(r_3)} = -63.3$ dB (W/m^2), leading to a reduction in the EMI of $R_{TR} = 2.8$ dB.

5.4.2 Statistical Analysis

We recall that the experimental measurements were conducted in three houses in France. For each Tx–Rx outlet configuration, the PLC signal was injected onto the three possible ports allowed in an MIMO communication: L–N, L–PE, and N–PE. The radiation level was evaluated at different locations around the electrical network. The following statistical analysis is based on 54 couples of channel measurements $H(f, r_0)$ and EMI measurements $H(f, r_3)$.

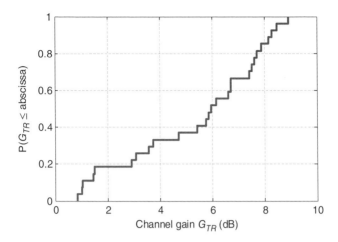

Figure 5.6 Cumulative distribution of the experimental channel gain G_{TR}.

We first considered the gain in total received power due to the application of TR filtering, G_{TR}, as defined in Equation (5.8). Figure 5.6 presents the cumulative distribution function (CDF) of this parameter. Interestingly, the application of TR always provides a positive gain comprised between 0.8 dB and 8.9 dB for our measurement database. Such an increase in the total received power has already been observed in the case of radio propagation with the same order of magnitude [19].The median channel gain is measured at 6 dB, and in 78% of the cases, the channel gain is higher than 3 dB.

Note that this channel gain could be exploited in different ways at the system level. With the objective of optimizing the system performance, such power gain at the Rx will directly lead to an increase in the Rx signal-to-noise ratio, and thus the bit error rate (BER) will be reduced. More interestingly, in our specific PLC application, the application of TR may also target a reduction in the undesired radiated power, while maintaining the same level of transmission performance. To achieve this goal, the Tx power can be reduced by a factor G_{TR}, leading to a reduction of the EMI by the same factor.

Figure 5.7 represents the CDF of the EMI reduction coefficient R_{TR} obtained in our measurement campaign. One observes that

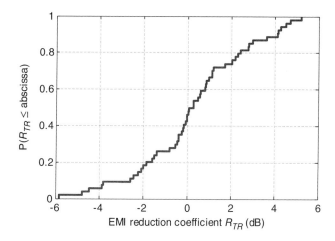

Figure 5.7 Cumulative distribution of the experimental EMI reduction coefficient R_{TR}.

in 54% of the cases, applying TR filtering allows a direct reduction of the radiated power, up to a maximum of 5.2 dB. However, in 46% of the cases, the application of TR leads to an increase in the EMI. In the worst case, a 5.9 dB increase factor is observed. This trend was already observed on a similar analysis performed on a series of measurements realized in the time domain [25]. When looking at specific cases, it can be concluded that mitigation of the EMI is more effective when the correlation between the channel CTF $H(f, r_0)$ and the EMI CTF $H(f, r_3)$ is low. The probability of strong decorrelation is higher when the topology of the electrical network is complex. Indeed, this results in a rich combination of multiple paths with a large variety of frequency fading patterns at different Rx locations.

Using the statistical analysis of the channel gain G_{TR} and of the EMI reduction coefficient R_{TR} as inputs, it is possible to devise an optimal strategy for an effective EMI reduction in the PLC context. As explained above, the channel gain G_{TR} obtained through the application of TR filtering allows the injected power level at the Tx to be reduced without modifying the system performance. However, when using this power backoff strategy, the EMI reduction coefficient R_{TR} still applies. Finally, the effective

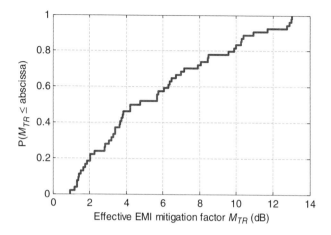

Figure 5.8 Cumulative distribution of the experimental effective EMI mitigation, M_{TR}.

EMI mitigation factor is given as the sum of the power backoff and the EMI reduction coefficient:

$$M_{TR} = G_{TR} + R_{TR} \qquad (5.14)$$

Figure 5.8 represents the CDF of the effective EMI mitigation factor M_{TR}, obtained from the set of experimental measurements mentioned above. The first observation is that in all investigated cases, the effective EMI mitigation factor is always positive and greater than 1 dB. This shows that the use of TR filtering associated with the optimal mitigation strategy is efficient in reducing the EMI generated by PLC systems.

More precisely, in 71% of the observed cases, the EMI is reduced by more than 3 dB. The mitigation factor can increase by up to 13 dB in the most favorable situations. Such cases correspond to configurations where the correlation between the channel CTF and the EMI CTF is particularly low. These results confirm a similar trend that we have already observed from measurements taken in the time domain in a laboratory environment [25]. Thus, the application of the TR technology to a PLC seems a promising method to mitigate undesired EMI radiating from the electrical network.

5.5 Conclusion

In this chapter, we presented a technique for the mitigation of radiated emissions from a PLC network by applying TR filtering to the Tx modem. TR is a technique first developed for acoustic or wireless electromagnetic waves, with the property to focus the received signal at the intended Rx location, both in the time and spatial domains. The application of TR to a wireline communication system was initially presented and validated using a limited set of measurements performed in the time domain [25]. These first results demonstrated that the focusing properties of TR could be exploited for wireline communications, which resulted in a decrease of the observed radiation.

In the present study, we applied the method to a series of experimental measurements collected in three houses in France using frequency-domain equipment. In addition, three modes of differential signal injection, using the live, neutral, and protective earth wires, were tested, as for current MIMO PLC modems. In total, 54 couples of channel and EMI measurements observed in the frequency range from 2 MHz to 28 MHz were used in our statistical study.

Results showed that the application of TR to PLC transmission over an electrical network allowed a gain in total received power varying between 0.8 dB and 8.9 dB. Such gains had already been observed in experiments conducted on wireless fields. Simultaneously, the total electromagnetic radiation of the electrical network is reduced in 54% of the cases, with a maximal reduction of 5.2 dB. By combining these two advantages, it is possible to transmit data with the same performance in terms of BER, while significantly reducing the electromagnetic radiation. For all tested cases in our study, the total EMI could be reduced. In 71% of the cases, the effective EMI reduction factor was greater than 3 dB, with a maximum mitigation factor of 13 dB. As a future work, this mitigation technique could be tested using a larger database of in-home PLC measurements in the broadband PLC frequency range, for instance covering different countries with different wiring practices. In addition, the same technique could be investigated for other technologies, such as the narrowband PLC techniques used for outdoor Smart Grids in the frequencies below 500 kHz, or the DSL technologies for the transmission over telephone lines.

References

1 L. T. Berger, A. Schwager, P. Pagani, and D. M. Schneider, eds, *MIMO Power Line Communications: Narrow and Broadband Standards, EMC, and Advanced Processing*, Boca Raton: CRC Press, 2014.

2 IEEE 1901–2010, IEEE Standard for Broadband over Power Line Networks: Medium Access Control and Physical Layer Specifications, December 2010.

3 ITU-T G.9960, Unified High-Speed Wireline-Based Home Networking Transceivers – System Architecture and Physical Layer Specification, June 2010.

4 HomePlug Alliance, HomePlug AV2 Specification, version 2.0, January 2012.

5 S. Galli, A. Scaglione, and Z. Wang, "For the grid and through the grid: the role of power line communications in the Smart Grid," *Proceedings of the IEEE*, vol. 99, no. 6, pp. 998–1027, June 2011.

6 V. Oksman and J. Zhang, "G.hnem: the new ITU-T Standard on narrowband PLC technology," *IEEE Communications Magazine*, vol. 49, no. 12, pp. 36–44, December 2011.

7 M. Ishihara, D. Umehara, and Y. Morihiro, "The correlation between radiated emissions and power line network components on indoor power line communications," in *IEEE International Symposium on Power Line Communications*, Orlando, Florida, March 2006.

8 Seventh Framework Programme: Theme 3 ICT-213311 OMEGA, Deliverable D3.3, "Report on Electro Magnetic Compatibility of Power Line Communications," December 2009.

9 A. Schwager, W. Bäschlin, *et al.*, "European MIMO PLC field measurements: overview of the ETSI STF410 campaign and EMI analysis," in *IEEE International Symposium on Power Line Communications and Its Applications (ISPLC)*, pp. 304–309, Beijing, China, March 2012.

10 Federal Communications Commission, Title 47 of the Code of Federal Regulations Part 15, 2007.

11 CENELEC, Final Draft of European Standard FprEN 50561-1, "Power Line Communication Apparatus Used in Low Voltage Installations – Radio Disturbance Characteristics – Limits and Methods of Measurement – Part 1: Apparatus for In-Home Use," June 2011.

12 A. Vukicevic, M. Rubinstein, F. Rachidi, and J-L. Bermudez, "On the impact of mitigating radiated emissions on the capacity of PLC systems," in *IEEE International Symposium on Power Line Communications and Its Applications*, pp. 487–492, Pisa, Italy, March 2007.

13 P. Favre, C. Candolfi, and P. Krahenbuehl, "Radiation and disturbance mitigation in PLC networks", in *20th International Zurich Symposium on Electromagnetic Compatibility*, pp. 5–8, Zürich, Switzerland, January 2009.

14 G. Lerosey, J. de Rosny, A. Tourin, A. Derode, G. Montaldo, and M. Fink, "Time reversal of electromagnetic waves," *Physical Review Letters*, vol. 92, pp. 1939041–1939043, May 2004.

15 A. E. Akogun, R. C. Qiu, and N. Guo, "Demonstrating time reversal in ultra-wideband communications using time domain measurements," in *51st International Instrumentation Symposium*, Knoxville, TN, May 2005.

16 A. Khaleghi, G. El Zein, and I. H. Naqvi, "Demonstration of time-reversal in indoor ultra-wideband communication: time domain measurement," in *4th International Symposium on Wireless Communication Systems*, pp. 465–468, Trondheim, Norway, October 2007.

17 A. Derode, P. Roux, and M. Fink, "Acoustic time-reversal through high-order multiple scattering," in *Proceedings of the IEEE Ultrasonics Symposium*, vol. 2, pp. 1091–1094, Seattle, WA, November 1995.

18 D. R. Jackson and D. R. Dowling, "Phase conjugation in underwater acoustics," *Journal of the Acoustical Society of America*, vol. 89, pp. 171–181, January 1991.

19 P. Pajusco and P. Pagani, "On the use of uniform circular arrays for characterizing UWB time reversal," *IEEE Transactions on Antennas and Propagation*, vol. 57, no. 1, pp. 102–109, January 2009.

20 C. Zhou and R. C. Qiu, "Spatial focusing of time-reversed UWB electromagnetic waves in a hallway environment," in *Southeastern Symposium on System Theory*, pp. 318–322, Cookeville, TN, March 2006.

21 M. Tlich, A. Zeddam, F. Moulin, and F. Gauthier, "Indoor power line communications channel characterization up to 100 MHz – Part I: One-parameter deterministic model," *IEEE Transactions on Power Delivery*, vol. 23, no. 3, pp. 1392–1401, July 2008.

22 M. Tlich, A. Zeddam, F. Moulin, and F. Gauthier, "Indoor power line communications channel characterization up to 100 MHz – Part II: Time-frequency analysis," *IEEE Transactions on Power Delivery*, vol. 23, no. 3, pp. 1402–1409, July 2008.

23 A. M. Tonello and F. Versolatto, "Bottom-up statistical PLC channel modeling – Part I: Random topology model and efficient transfer function computation," *IEEE Transactions on Power Delivery*, vol. 26, no. 2, pp. 891–898, April 2011.

24 A. M. Tonello and F. Versolatto, "Bottom-up statistical PLC channel modeling – Part II: Inferring the statistics," *IEEE Transactions on Power Delivery*, vol. 25, no. 4, pp. 2356–2363, October 2010.

25 A. Mescco, P. Pagani, M. Ney, and A. Zeddam, "Radiation mitigation for power line communications using time reversal," *Journal of Electrical and Computer Engineering*, vol. 2013, paper ID 402514, March 2013.

26 ETSI TR 101 562-1 V1.3.1 Technical Report, Powerline Telecommunications (PLT), MIMO PLT, Part 1: Measurements Methods of MIMO PLT, Chapter 7.1, 2012.

6

Application of Electromagnetic Time Reversal to Lightning Location

M. Rubinstein[1] and F. Rachidi[2]

[1] *University of Applied Sciences of Western Switzerland, Yverdon, Switzerland*
[2] *Swiss Federal Institute of Technology (EPFL), Lausanne, Switzerland*

6.1 Introduction

Lightning is a major source of electromagnetic interference and damage to electronic circuits, buildings, and other exposed man-made structures such as wind turbines and photovoltaics [1, 2]. Lightning is also responsible for numerous forest fires and for the death of livestock and humans.

Systems for the detection and location of lightning have been used for many decades to record lightning activity [3]. Many techniques have been proposed to locate lightning, including the use of the magnetic field to determine the direction to the lightning channel (e.g., [4]), the use of the time of arrival of electromagnetic radiation at multiple sensors (e.g., [5]), interferometric direction finding (e.g., [6]), the use of acoustic information from thunder to map the lightning channel (e.g., [7]), a combination of time-of-arrival and peak amplitude of the radiated electromagnetic fields (e.g., [8]), and even seismic waves [9]. For a good historical review of the development of lightning location systems, see [10].

Electromagnetic Time Reversal: Application to Electromagnetic Compatibility and Power Systems,
First Edition. Edited by Farhad Rachidi, Marcos Rubinstein and Mario Paolone.
© 2017 John Wiley & Sons, Ltd. Published 2017 by John Wiley & Sons, Ltd.
Companion Website: www.wiley.com/go/rachidi55

Most commercial lightning location networks nowadays, however, are based on two of these techniques, used either separately or in combination through appropriate algorithms: the magnetic direction finding (MDF) technique and the time-of-arrival (ToA) technique, also known as difference in time-of-arrival (DToA).

In this chapter, we will first describe, in Section 6.2, the main processes in cloud-to-ground lightning, since this knowledge and the associated terminology are important to understand lightning location techniques. We will then review the principles used by the main classical location techniques that will be compared to time reversal at the end of Section 6.3. In Section 6.3, we will present the electromagnetic time-reversal (EMTR) lightning location technique and then give both a mathematical proof and verification of the technique by simulations. The important issue of the application of EMTR in the presence of a finitely conducting ground will also be dealt with in that section. At the end of the section, we will discuss the relation between EMTR and other lightning location techniques. The last section of the chapter is dedicated to practical implementation issues.

6.2 Overview of Lightning Location Techniques

6.2.1 Cloud-to-Ground Lightning[1]

Lightning is a long electric discharge through the air that involves charge centers in the cloud, on the Earth's surface (or objects on it) and in the upper atmosphere. A complete lightning is called a lightning flash. Flashes whose endpoints are in the cloud and the ground are called cloud-to-ground flashes or ground flashes. A lightning flash between charge centers in the cloud, not involving the ground, is called a cloud flash. A third type of lightning that occurs between cloud tops and the ionosphere is known collectively as transient luminous events.

Although cloud-to-ground lightning represents only about one-fourth of all lightning, it is responsible for most effects on

---

1 For a more comprehensive description of lightning discharge, refer, for instance, to [3] and [11].

humans and human activity and much effort has been devoted to its study.

Four categories of cloud-to-ground lightning strikes were identified by Berger *et al.* [12] depending on the direction of the motion of the initial leader (upward or downward) and the sign of the charge transferred to ground (positive or negative). About 90% or more of global cloud-to-ground lightning are downward negative and 10% or less downward positive. Upward flashes occur either from tall structures or from objects of moderate height located on mountaintops.

In what follows, we will briefly describe the main processes in downward negative cloud-to-ground lightning.

The sources of cloud-to-ground lightning are charge centers in the clouds. Cloud electrification is not fully understood. A proposed theory involves charge transfer between colliding graupel (a form of sleet) and ice crystals at the appropriate temperature range in the clouds. Another proposed mechanism involves convection of charges generated by corona and cosmic rays. The resulting charge structure can be roughly approximated by a set of three charge centers (the so-called tripole model), a positive near the top of the thundercloud, with a net charge of a few tens of coulombs, a negative charge center below it, with a net charge equal to the positive charge above it, and a small positive charge near the bottom of the cloud of a few coulombs.

Downward negative cloud-to-ground lightning starts in the cloud with a process known as the *preliminary discharge*, believed to happen between the lower positive charge and the main negative charge in the cloud. The preliminary breakdown is followed by a faint discharge that emerges from the bottom of the cloud and progresses towards the ground in a stepped manner. This second process is known as *the stepped leader* and it advances at average speeds on the order of 100 km/s. As it progresses towards the ground, the stepped leader branches downward, taking the shape that could be described as that of an upside-down tree. When the stepped leader approaches the ground, upward discharges develop from pointed objects in the ground and one of these makes contact with one of the branches in what is known as the *attachment process*. The attachment process gives rise to a strong, impulsive current that travels up the plasma channel of the stepped leader at speeds on the order of

100 000 to 200 000 km/s. This very fast and energetic process is known as the *first return stroke*. Only about 1 out of every 7 negative cloud-to-ground flashes ends at this point. In the remaining flashes, the leader/return-stroke sequence repeats itself a number of times, with pauses typically lasting tens of milliseconds between successive leader/return-stroke sequences. The subsequent leader, known as a *dart leader*, is fundamentally different from the stepped leader in that it develops continuously, without stepping, towards the ground and its average speed is some 10 times higher than that of the stepped leader. *Subsequent return strokes* are characterized, on average, by smaller peak currents and faster risetimes compared to first return strokes. The average number of leader/return stroke sequences in a cloud-to-ground flash is about 4.

6.2.2   Magnetic Direction Finding

The first investigations on the magnetic direction finding lightning location technique were reported some sixty years ago by Horner [13, 14]. The technique uses sensors capable of determining the direction to the lightning based on the fact that, at ground level, the radiation component of the magnetic field from the lightning return stroke current is perpendicular to the direction of propagation of the electromagnetic wavefront. The strike point is determined as the intersection of the lines from two of the sensors in the direction of the lightning, as illustrated in Figure 6.1, in which the magnetic field angles are measured with respect to North. Although only two sensors are shown in Figure 6.1, a minimum of three sensors is required to cover all azimuth angles, since two sensors cannot locate lightning that strikes along the straight line that passes through them.

It can be readily shown that knowledge of the positions of the two sensors and of the angles from each sensor to the lightning strike point is sufficient to calculate the coordinates of the strike location through simple trigonometry.

6.2.3   Difference in Time of Arrival (DToA)

The difference in time-of-arrival technique is based on the detection of the time of arrival of the electromagnetic pulse generated

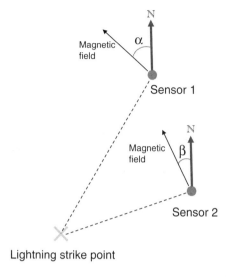

Figure 6.1 Illustration of the principle of direction finding based on two sensors for a flat ground. The magnetic field is perpendicular to the direction of propagation of the electromagnetic wavefront coming from the lightning. The angles are measured relative to North.

by the return stroke current at a minimum of four sensors. The difference in time of arrival at each pair of sensors defines two branches of a hyperbola. It is thus possible to determine the strike point from the intersection of these curves, as illustrated in Figure 6.2 where, for clarity of presentation, the hyperbolic branches associated with the fourth sensor, required to solve the ambiguity created by the two intersection points, are not shown.

6.3 EMTR and Lightning Location

Recently, electromagnetic time reversal (EMTR) has been proposed as a technique to locate lightning discharges with potentially higher accuracy than classical techniques [15]. In the next sections, we will describe the way in which EMTR can be used for lightning location and we will discuss some of the issues

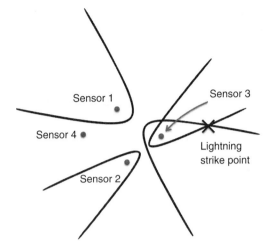

Figure 6.2 Illustration of the principle of the difference in time of arrival. The hyperbolic branches associated with the fourth sensor were left out for clarity.

associated with the practical implementation of the technique in real systems.

6.3.1 Description of the Method

EMTR can be used to locate the position of sources of electromagnetic radiation in lightning. Here we will show how the strike point of a cloud-to-ground flash can be located, but the same technique can be implemented for the location of impulsive currents in other lightning processes. The method is based on the simultaneous, synchronized recording of the electric or the magnetic field waveforms at several sensors. The records of the waveforms are then transmitted to a central server that time-reverses them and lets the time-reversed versions back-propagate into the location domain by numerical simulation. As we will show in Section 6.3.2.1, the back-propagated fields will add up in phase at the lightning strike location. If the attenuation is artificially removed from the back-propagation model, the peak value of the sum of the back-propagating fields will be maximum at the location of the strike point since it is there that maximum constructive interference occurs. To understand the method better, let us refer to Figure 6.3.

6.3.2 Principle of Operation

In what follows, we will present a mathematical proof of the EMTR technique for lightning location and a verification of the technique based on simulations.

6.3.2.1 The Mathematical Proof

A theoretical justification of the EMTR lightning location method was presented in [16], on which the following derivation is based.

We will first introduce the notation to be used and we will then write an equation for the total time-reversed field injected by the sensors, as shown in Figure 6.3c, into the location domain. We will then show that these fields interfere constructively to produce an overall maximum at the location of the original electromagnetic source.

With reference to Figure 6.4, we will call the position vector of the lightning strike point r_s, and the position vectors of N sensors r_n, where n is an integer that identifies the sensor and therefore ranges from 1 to N.

Assuming that the cloud-to-ground lightning channel is straight and vertical and that the ground is flat and perfectly conducting (we will relax the latter assumption later on), the far magnetic field at ground level from the lightning return stroke current is inversely proportional to the distance from the lightning to the observation point. This relation can be written in the time domain as

$$H(r, t) = \frac{w\left(t - \frac{|r - r_s|}{c}\right)}{|r - r_s|} a_\phi \tag{6.1}$$

where $w(t)$ is a function whose form depends on the behavior of the current as a function of the time and the height along the lightning channel. Although $w(t)$ could be estimated from measurements and from an independent determination of the position of the lightning strike point using (6.1), its form can, and usually is, obtained from so-called return-stroke models that

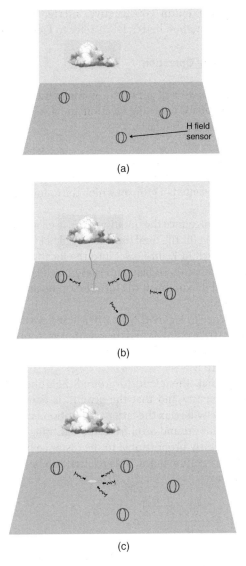

(a)

(b)

(c)

Figure 6.3 Illustration of EMTR for lightning location. (a) Four crossed-loop magnetic field sensors. (b) Lightning strikes and the impulsive radiation from one of its return strokes reaches the sensors. (c) The time-reversed versions of the field waveforms are back-propagated into the location domain by numerical simulation to find the point of maximum constructive interference.

Figure 6.4 *N* magnetic field sensors and the lightning strike point.

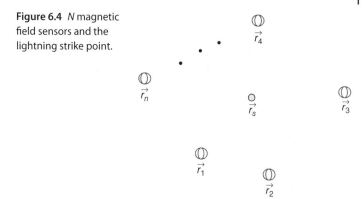

have been proposed in the literature (see, for instance, [17] and [18]).[2] In (6.1), a_ϕ is the unit vector in the ϕ direction in a cylindrical (or spherical) coordinate system.

Evaluating (6.1) at r_n, we obtain the field measured by sensor n,

$$H(r_n, t) = \frac{w\left(t - \frac{|r_n - r_s|}{c}\right)}{|r_n - r_s|} a_\phi \tag{6.2}$$

As explained in Section 6.3.1, each sensor calculates a time-reversed version of the measured field that will then be retransmitted into the location domain. The time-reversed fields can be calculated from (6.2) by replacing t by $T - t$, where the constant shift T, which is greater than or equal to the duration of the measured field, is added to adjust the reference time in such a way that the time-reversed waveforms start at $t \geq 0$ and not at a negative time. The time-reversed fields are thus given by

$$H_{TRn}(t) = \frac{w\left(T - t - \frac{|r_n - r_s|}{c}\right)}{|r_n - r_s|} a_\phi \tag{6.3}$$

where the subindex TRn has been added to the field to indicate that it is no longer a field measured at the nth sensor location but rather the time-reversed field calculated by that sensor.

---

2 Note that while (6.1) can be used to calculate the magnetic field, much of the scientific and engineering literature on return stroke modeling uses the expression for the vertical electric field instead, which can be obtained by multiplying (1) by the field impedance, 120π, assuming far-field conditions.

At this point, the sensors, following the EMTR location method, re-inject the time-reversed fields in (6.3) into the medium by numerical simulation and they let it propagate removing the amplitude dependence on the distance.[3] Under these conditions, the field from the nth sensor at an arbitrary position r is given by

$$H_{TR}(t) = \frac{w\left(T - t - \frac{|r_n - r_s|}{c} + \frac{|r_n - r|}{c}\right)}{|r_n - r_s|}a_\phi \qquad (6.4)$$

where we have added the time shift introduced by propagation from the sensor to the field point r.

The total field at an arbitrary position can now be calculated by adding the contributions from the N sensors:

$$H_{TRtotal}(t) = \sum_{n=1}^{N} \frac{w\left(T - t - \frac{|r_n - r_s|}{c} + \frac{|r_n - r|}{c}\right)}{|r_n - r_s|}a_\phi \qquad (6.5)$$

The terms in the summation in (6.5) have the same waveshape (given by the function w) and they differ only in their relative time shift, which comes from the terms accompanying t in the argument.

The maximum amplitude field will be obtained when all of the terms contributing to the total field in (6.5) are in phase. By inspection, it can be readily seen that this condition is satisfied at the strike point, where $r = r_s$. Thus, at the strike point, Equation (6.5) becomes

$$H_{TRtotal}(t) = w(T - t)a_\phi \sum_{n=1}^{N} \frac{1}{|r_n - r_s|} \qquad (6.6)$$

6.3.2.2 Simulations

In this section, simulations are used to validate the electromagnetic time reversal lightning location method following the procedure presented by Mora *et al.* [15].

---

3 It is also possible to work with the amplitude dependence, in which case the identification of the lightning strike point is based only on the cross-correlation of the different waveforms and not on the maximum amplitude point. Ideally, both criteria should be used for the location. It is also worth noting that the $1/r$ amplitude dependence in the back propagation might result in prohibitively small fields beyond some critical distances, leading to numerical errors [16].

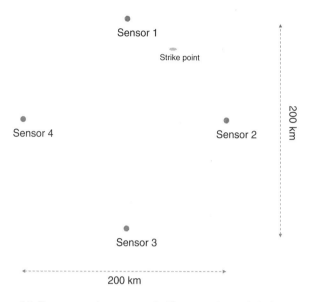

Figure 6.5 Four sensors in a symmetrical layout and a single lightning strike point used in the simulation.

A simulation domain with four sensors and a lightning strike point is shown in Figure 6.5. A simulation of the lightning strike is carried out to calculate the magnetic fields that will be measured by the sensors. The fields from lightning return strokes can be calculated using any of a number of return stroke models that have been proposed in the literature (reviews on return stroke modeling can be found in [19] and [20]). In the simulations in this section, the transmission line model [21] will be used due to its simplicity.

The return stroke channel base current that was used as an input to the transmission line return stroke model, represented by the sum of two Heidler functions [22], is of the form

$$i(t) = \frac{I_{01}}{\eta_1} \frac{\left(\frac{t}{\tau_{11}}\right)^{n_1}}{1 + \left(\frac{t}{\tau_{11}}\right)^{n}} e^{-t/\tau_{21}} + \frac{I_{02}}{\eta_2} \frac{\left(\frac{t}{\tau_{12}}\right)^{n}}{1 + \left(\frac{t}{\tau_{12}}\right)^{n}} e^{-t/\tau_{22}}$$

(6.7)

with I_{01} = 13 kA, τ_{11} = 0.15 μs, τ_{21} = 3 μs, n_1 = 2, η_1 = 0.73, I_{02} = 7 kA, τ_{21} = 5 μs, τ_{22} = 50 μs, n_2 = 8, and η_2 = 0.86.

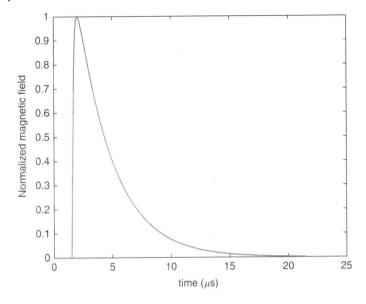

Figure 6.6 Normalized magnetic field.

The return stroke radiation fields at ground level have the same waveshape as that of the channel base current. The normalized magnetic radiation field is shown in Figure 6.6. This represents the field measured by each sensor.

This received field is time-reversed in preparation for back-injection into the location domain. The time-reversed field corresponding to the magnetic field waveform in Figure 6.6 is shown in Figure 6.7.

The time-reversed field in Figure 6.7 was back-propagated from each sensor into the location domain and the total field was observed at every point in the simulation domain grid.

Figure 6.8 shows the total back-propagated, time-reversed fields from the simulations at three different positions in the simulation domain. The total field in Figure 6.8a was calculated at a point that is 10 km away from the actual strike point. The back-propagated contributions from the four sensors are clearly distinguishable in that case and the amplitude is much lower than that at the strike point. The overall waveshape is clearly different from that in Figure 6.7. The waveform in Figure 6.8b represents the total back-propagated field 1 km away from the lightning

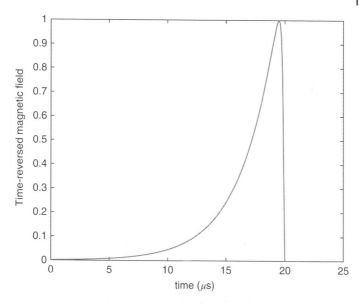

Figure 6.7 Normalized time-reversed magnetic field.

strike point. The contributions from the four back-propagated waveforms are no longer discernible but the waveform is still clearly different from that in Figure 6.7. In addition, the amplitude is about half of that at the strike point. Finally, the field at the actual location of the strike point (Figure 6.8c) has the highest amplitude and its shape is that of the time-reversed field at the sensors. Note that the amplitude is 4 since it is the in-phase addition of the normalized fields from the four sensors.

An additional graphical representation of a forward propagating impulsive wave from a lightning at the center of the simulation domain, with 14 sensors around a strike point, can be seen in Figure 6.9a. The wave is seen to propagate out from the source towards the sensors, located near the sides of the domain. In Figure 6.9b, the back-propagated, time-reversed waves from the different sensors focus at the strike point.

6.3.3 Finite Ground Conductivity

In the previous sections, we have assumed that propagation occurs without losses since the ground is a perfect conductor.

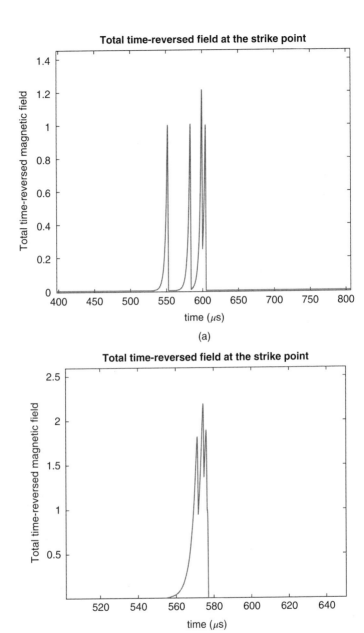

Figure 6.8 Simulated total time-reversed fields. (a) Total TR magnetic field at a point 10 km from the actual strike point. (b) Total TR magnetic field at a point 1 km away from the strike point. (c) Total magnetic field at the strike point.

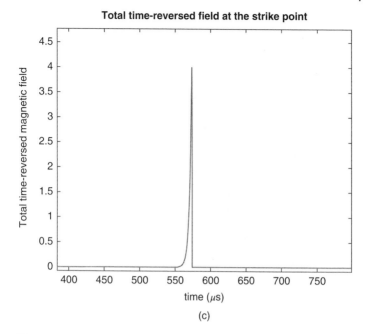

Figure 6.8 (*Continued*)

This allowed us to use Maxwell's equations in a form that, as discussed in Chapter 1, is time-reversal invariant. In this section, we discuss the case of a finitely conducting ground. In a real EMTR location system, the fields measured by the sensors (illustrated in Figure 6.3b and shown in the simulated waveform in Figure 6.6) would be distorted by propagation over the imperfectly conducting ground. The effect of losses on fields propagating over an imperfect ground with a constant conductivity can be accounted for by way of the following convolutive factor [23]:

$$S_f(|\boldsymbol{r} - \boldsymbol{r}_s|, t) = \frac{\mathrm{d}}{\mathrm{d}t} \left[1 - \mathrm{e}^{-t^2/4\zeta^2} + 2\beta(\varepsilon_r + 1)\frac{J(x)}{t} \right] \quad (6.8)$$

in which $x = t/2\zeta$, $\zeta^2 = |\boldsymbol{r} - \boldsymbol{r}_s|/2\mu_0\sigma c^3$, μ_0 is the permeability of vacuum, ε_r is the relative permittivity of the ground, σ is the conductivity of the ground, c is the speed of light, and $J(x) = x^2(1 - x^2)\mathrm{e}^{-x^2}$, $\beta = 1/\mu_0\sigma c^2$.

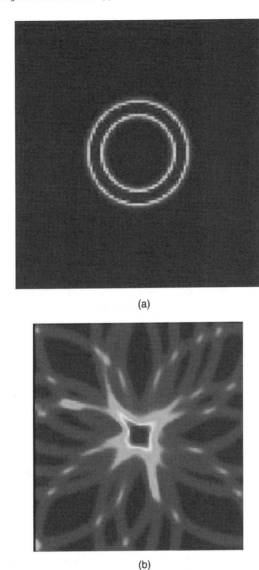

(a)

(b)

Figure 6.9 Graphical representation of (a) a wave propagating out from a lightning strike at the center of the simulation domain and (b) the focusing effect of the back-propagating, time-reversed wavefronts retransmitted by the sensors. An animated version of this figure can be found www.wiley.com/go/rachidi55.

Equation (6.2) can therefore be written for the case of a finitely conducting ground as

$$H_L(r_n, t) = \frac{w\left(t - \frac{|r_n - r_s|}{c}\right)}{|r_n - r_s|} \otimes S_f(|r_n - r_s|, t)a_\phi \qquad (6.9)$$

where $\otimes$ is the convolution operator and the subindex L is used to indicate that losses were taken into account.

We will study the effect of the ground conductivity in the frequency domain since the equations can be written in a simpler manner.

Equation (6.9) can be readily transformed into the frequency domain using the properties of the Fourier Transform:

$$\underline{H}_L(r_n, t) = \frac{\underline{w}(\omega)}{|r_n - r_s|}\underline{S}_f(|r_n - r_s|, \omega)e^{-j\omega\frac{|r_n - r_s|}{c}}a_\phi \qquad (6.10)$$

where ω is the angular frequency and where an underline has been used to distinguish the complex variable functions in (6.10) from the real-valued functions in (6.9). In (6.10), the exponential accounts for the propagation phase shift from the source to the nth sensor.

We will refer henceforth to the frequency-domain factor $\underline{S}_f(|r_n - r_s|, \omega)$, which accounts for the effect of the propagation over a finite-conductivity ground, as the finite-conductivity transfer function.

It can be shown that the time-reversal operation in the time domain corresponds to the conjugate operation in the frequency domain. Thus, we can write the time-reversed version of the field at the nth sensor by taking the conjugate of (6.10):

$$\underline{H}_{TR\_L}^*(r_n, t) = \frac{\underline{w}^*(\omega)}{|r_n - r_s|}\underline{S}_f^*(|r_n - r_s|, \omega)e^{j\omega\frac{|r_n - r_s|}{c}}a_\phi \qquad (6.11)$$

Equation (6.11) represents the time-reversed field that will be back-propagated to obtain a fix for the lightning strike point when losses are considered.

Three different back-propagation models were proposed and compared in [16] and we will describe each of them in the following three sections.

6.3.3.1 Perfect Ground Back-Propagation Model

In this first back-propagation model to be used in the simulations, the ground is assumed to be a perfect conductor. Therefore, the back-propagation will only introduce a phase shift equal to $e^{-j\omega\frac{|r_n-r|}{c}}$. The back-propagated field at position r is therefore given by

$$\underline{H}^*_{TR\_L}(r,t) = \frac{w^*(\omega)}{|r_n-r_s|}\underline{S}^*_{\underline{f}}(|r_n-r_s|,\omega)e^{j\omega\frac{|r_n-r_s|-|r_n-r|}{c}}a_\phi$$

(6.12)

Equation (6.12) was obtained by adding the phase shift to the exponential factor in (6.11).

At the lightning strike location, the field equals

$$\underline{H}^*_{TR\_L}(r_s,t) = \frac{w^*(\omega)}{|r_n-r_s|}\underline{S}^*_{\underline{f}}(|r_n-r_s|,\omega)a_\phi$$

(6.13)

To compare this back-propagation model with the case studied in Section 6.3.2.1, in which a perfectly conducting ground was assumed for the forward- and for the back-propagation, we write here the frequency-domain version of (6.4) evaluated at the lightning strike location that we wish to compare to (6.13):

$$\underline{H}^*_{TR}(r_s,t) = \frac{w^*(\omega)}{|r_n-r_s|}a_\phi$$

(6.14)

The difference between (6.13) and (6.14) is, as expected, the presence in (6.13) of the finite-conductivity transfer function $\underline{S}_{\underline{f}}$, which introduces dispersion in the perfectly conducting ground back-propagation model. The distortion appears both in the amplitude and in the phase. As a consequence, the contributions from the different sensors will no longer perfectly add in phase at the strike point.

6.3.3.2 Lossy Ground Back-Propagation Model

The lossy ground back-propagation model lets the fields back-propagate under the same imperfectly conducting ground conditions as the forward-propagation. The back-propagation will therefore include, in addition to the phase shift, further dispersion due to the inclusion of the finite-conductivity transfer function $\underline{S}_{\underline{f}}$ in the back-propagation. Under this model, the back-propagated field at position r is given by the multiplication of (6.11) and the finite-conductivity transfer function:

$$\underline{H}^*_{TR\_L}(r, t) = \frac{w^*(\omega)}{|r_n - r_s|} \underline{S}^*_f(|r_n - r_s|, \omega)$$

$$\times \underline{S}_f(|r_n - r|, \omega) e^{j\omega \frac{|r_n - r_s| - |r_n - r|}{c}} a_\phi \qquad (6.15)$$

At the source location, the field from the nth sensor is given by

$$\underline{H}^*_{TR\_L}(r_n, t) = \frac{w^*(\omega)}{|r_n - r_s|} |\underline{S}_f(|r_n - r_s|, \omega)|^2 a_\phi \qquad (6.16)$$

where we have used the fact that the product of the finite-conductivity transfer function and its conjugate equals the square of the magnitude of the function.

A comparison of the field at the lightning location point corresponding to this model and given by (6.16) to the field in the case of perfect forward- and back-propagation given by (6.14) reveals that the only difference is the multiplicative real, frequency-dependent factor $|\underline{S}_f(|r_n - r_s|, \omega)|^2$. Therefore, unlike the perfect ground back-propagation model, only the amplitude is distorted by the lossy ground back-propagation model. Indeed, the phase dispersion introduced by the forward propagation is compensated by the time-reversal and the back-propagation. The back-propagated fields from the different sensors will, in general, exhibit different waveshapes under the lossy ground back-propagation model since the multiplicative finite-conductivity transfer function is both frequency dependent and distance dependent.

6.3.3.3 Inverted-Loss Back-Propagation Model

The inverted-loss back-propagation model uses the back-propagation to compensate for the dispersion introduced in the forward-propagation. The end result is a field at the lightning strike location that exhibits neither phase nor amplitude distortion. To achieve this, the inverted-loss back-propagation model applies the equalization filter $1/|\underline{S}^*_f(|r_n - r|, \omega)|$ to the forward-propagated fields. The back-propagated field at position r is therefore given by

$$\underline{H}^*_{TR\_L}(r, t) = \frac{\underline{w}^*(\omega)}{|r_n - r_s|} \underline{S}^*_f(|r_n - r_s|, \omega)$$

$$\times \frac{1}{\underline{S}^*_f(|r_n - r|, \omega)} e^{j\omega \frac{|r_n - r_s| - |r_n - r|}{c}} a_\phi \quad (6.17)$$

Evaluating (6.17) at the strike location point r_s, the field is given by

$$\underline{H}^*_{TR\_L}(r_s, t) = \frac{\underline{w}^*(\omega)}{|r_n - r_s|} a_\phi \quad (6.18)$$

which is identical to the field for the perfectly conducting ground forward- and back-propagation case that we had calculated in (6.14).

The inverted-loss back-propagation model can be seen as a variation of the perfect ground back-propagation model in which the fields in (6.12) are divided by $\underline{S}^*_f(|r_n - r|, \omega)$. It can also be viewed as a variation on the lossy ground back-propagation model in which (6.15) is divided by $|\underline{S}_f(|r_n - r|, \omega)|^2$.

Unlike the first two models presented in the previous two sections, the inverted-loss back-propagation model completely eliminates the effect of the finite conductivity of the ground. However, as pointed out in [16], the practical application of the model requires precise knowledge of the electrical parameters of the ground.

Two more difficulties with this model were identified in [16]. The first one is the fact that $|\underline{S}_f(|r_n - r|, \omega)|$ tends to zero for high values of ω, which in turn leads to high amplitudes at higher frequencies in (6.17). This presents a problem for the inverse Fourier transform used to pass from the frequency to the time domain. Lugrin and co-workers [16] proposed the use of a zero-phase low-pass filter with an appropriate transfer function $\underline{H}(\omega)$ to address this problem. Modifying (6.17) to include this transfer function, one obtains

$$\underline{H}^*_{TR\_L\_F}(r, t) = \frac{\underline{w}^*(\omega)}{|r_n - r_s|} \frac{\underline{S}^*_f(|r_n - r_s|, \omega)}{\underline{S}^*_f(|r_n - r|, \omega)} \underline{H}(\omega) e^{j\omega \frac{|r_n - r_s| - |r_n - r|}{c}} a_\phi$$

$$(6.19)$$

where the subindex TR_L_F on the left-hand side signifies that the field is time-reversed, that it includes losses, and that it has been filtered.

The second difficulty lies in the fact that the signal amplitude increases with propagation distance and, therefore, the criterion of highest total peak amplitude to identify the strike point location is no longer applicable. Lugrin and co-workers proposed, as a possible solution, to normalize the filtered, back-propagated fields by dividing them by the peak time-domain amplitude at each position. The peak time-domain amplitude $A(|r_n - r|)$ to be used in the normalization can be calculated by applying the inverse Fourier transform to (6.19) and finding its peak amplitude:

$$
A(|r_n - r|)
$$
$$
= \max\left\{ F^{-1}\left[\frac{w^*(\omega)}{|r_n - r_s|} \frac{S_f^*(|r_n - r_s|, \omega)}{\underline{S}_f^*(|r_n - r|, \omega)} H(\omega) e^{j\omega \frac{|r_n - r_s| - |r_n - r|}{c}} a_\phi \right] \right\}
$$
$$
(6.20)
$$

in which F^{-1} denotes the inverse Fourier transform.

Modifying (6.19) by introducing into the filter the normalization factor, we get

$$
\underline{H}_{TR\_L\_F\_N}^*(r, t) = \frac{w^*(\omega)}{|r_n - r_s|} \frac{S_f^*(|r_n - r_s|, \omega)}{\underline{S}_f^*(|r_n - r|, \omega)}
$$
$$
\times \frac{H(\omega)}{A(|r_n - r|)} e^{j\omega \frac{|r_n - r_s| - |r_n - r|}{c}} a_\phi \quad (6.21)
$$

where the letter N at the end of the subindex on the left-hand side indicates that the field is normalized.

At the lightning strike point, (6.21) reduces to

$$
\underline{H}_{TR\_L\_F\_N}^*(r_s, t) = \frac{w^*(\omega)}{|r_n - r_s|} \frac{H(\omega)}{A(|r_n - r_s|)} a_\phi \quad (6.22)
$$

6.3.3.4   Comparison of the Back-Propagation Models

In Section 6.3.3.1, we saw that the time-reversed fields obtained by simulation with the perfect ground back-propagation model include both phase and amplitude distortion due to the uncompensated dispersion introduced by the finite ground conductivity during forward-propagation. On the other hand, the back-propagated fields obtained with the lossy ground back-propagation model presented in Section 6.3.3.2 should exhibit distortion only in the amplitudes of their frequency components since the phase distortion of the forward propagation is compensated by the back-propagation of the time-reversed fields. Finally, in the inverted-loss back-propagation model, both the amplitude distortion and the phase distortion are compensated. One would therefore expect the location accuracy of the inverted-loss back-propagation model to be the best and that of the perfect ground back-propagation model to be the worst of the three.

This was indeed the conclusion in [16] based on simulations for three different ground conductivities, 1 S/m, 0.01 S/m, and 0.002 S/m. Specifically, the location errors of the inverted-loss back-propagation model were found to be smaller than 10 m, which was the size of the cell in the simulation domain. For the perfect ground back-propagation model, the errors were of the order of several hundred meters. The lossy ground back-propagation model yielded location errors of over one hundred meters.

The inverted-loss back-propagation model is relatively insensitive to the value of the ground conductivity used in the back-propagation, with location errors of a few tens of meters if the conductivity is underestimated by a factor of two. At present, no studies have been presented on the influence of more complex grounds on the accuracy of lightning location by time reversal.

6.3.4 Relation between EMTR and Other Lightning Location Techniques

Although lightning location by EMTR is unrelated to direction-finding location techniques such as the MDF technique described in Section 6.2.2, Lugrin *et al.* [16] showed that it can be considered to be a special case of the DToA technique discussed in Section 6.2.3.

The proof is based on the EMTR constructive interference condition used to determine the lightning strike point in the EMTR technique. At the lightning strike point, each one of the time-reversed, back-propagating wavefronts must arrive with the same phase. Mathematically, we can set the time delay terms in Equation (6.4) equal to a constant K:

$$T - \frac{|r_n - r_s|}{c} + \frac{|r_n - r|}{c} = K \qquad (6.23)$$

Writing Equation (6.23) for two different sensors, i and j, we obtain

$$T - \frac{|r_i - r_s|}{c} + \frac{|r_i - r|}{c} = K \qquad (6.24)$$

$$T - \frac{|r_j - r_s|}{c} + \frac{|r_j - r|}{c} = K \qquad (6.25)$$

Subtracting (6.25) from (6.24), we get

$$-\frac{|r_i - r_s|}{c} + \frac{|r_i - r|}{c} + \frac{|r_j - r_s|}{c} - \frac{|r_j - r|}{c} = 0 \qquad (6.26)$$

Rearranging terms in (6.26), we can write that equation as

$$\frac{|r_j - r|}{c} - \frac{|r_i - r|}{c} = \frac{|r_j - r_s|}{c} + \frac{|r_i - r_s|}{c} \qquad (6.27)$$

The left-hand side in (6.27) is the difference in the times of arrival from a point r to the sensors i and j and the right-hand side is a constant equal to the difference in the times of arrival at sensors i and j. The equation therefore defines hyperbolic branches in the same way as the basic equation used in the difference in time of arrival technique.

Note that only the phase information, related to the propagation time, was used in this derivation. The EMTR technique includes the waveshape, in addition to the propagation time, in the determination of the strike point.

6.4 Practical Implementation Issues

In this section, we will discuss several issues associated with the practical implementation of the electromagnetic time-reversal lightning location technique.

6.4.1 Modeling of Complex Dissipative Medium

The location accuracy obtained with the time-reversal lightning location method depends on the quality of the back-propagation simulation model. The quality of the simulation depends on the accuracy of the simulation model itself and on the availability of the electrical and geometrical parameters of the ground. Knowledge of the ground parameters can be obtained by direct or indirect measurements and, given the current technological advances and trends, conductivity and ground structure atlases can be expected to improve over time. In addition, the accuracy of the simulation models is the object of intense research and it, too, can be safely expected to improve steadily in the years to come.

6.4.2 Simulation Resources

Lightning location by time reversal requires simulations of back-propagation over large simulation domains with appreciable complexity, involving different kinds of scatterers and a complex terrain profile. As discussed in Chapter 1, compared to a homogeneous medium, a higher focusing quality can be achieved in an inhomogeneous medium as a result of multiple reflections and scattering. Therefore, a more accurate simulation of the back-propagation phase taking into account terrain complexity would certainly result in higher performance of EMTR. Efficient codes running on fast computers with large memory resources are a requirement for real-time lightning location using this technique due to the needed numerical simulations. The increase in computing power, currently through multiple processor cores and the research on new materials, research on novel computing techniques, more efficient codes, and falling costs of computer hardware will lead to a reduction in simulation times and an increase in memory resources, enabling the use of complex models in EMTR back-propagation simulations.

6.4.3 Data Storage and Communication Resources

Although some of the deployed commercial lightning location networks record, at least temporarily, field waveforms for each lightning flash (e.g., GLD360 [24] and LINET [25]), others (e.g., EUCLID [26] and NLDN [10]) do not record them. The latter extract and store only the parameters of the waveforms that are required to locate the strike point and to infer important parameters such as the return stroke current peak. In the time-reversal lightning location technique, the full waveform is needed so that a time-reversed version of it can be back-propagated into the location domain.

A reason why only a few parameters are recorded for each lightning flash in any system is the lack of computer resources such as memory and disk space. Lugrin *et al.* [16] investigated the accuracy of the time-reversal lightning location technique if the full field waveforms are estimated using a limited number of parameters through extrapolation. The authors found that the extrapolation technique yielded results with insufficient accuracy.

The price of memory has been decreasing for decades and, as mentioned in the previous section, it is likely to continue to fall. In addition, the diminishing price of data communications will make it possible in the future to transmit and record the lightning field waveform data at an acceptable cost. On the other hand, memory and communication capacity requirements can also be reduced by way of advanced signal processing techniques to represent complex waveforms with only a few parameters as compared to the complete set of time samples. Karami *et al.* [27], for instance, have proposed the use of the matrix pencil method (MPM) (given in [28]) to express digitized field waveforms in terms of a reduced number of parameters and thus lower the computer memory required to store fields. An example from [27] is shown in Figure 6.10 in which a measured magnetic field

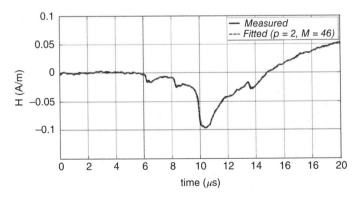

Figure 6.10 Lightning radiation magnetic field. The solid line represents the measured waveform and the dashed line is the fitted waveform using MPM. (Adapted from [27].)

waveform, which consisted of 10 000 points, was represented with only 46 poles. This represents a reduction of over two orders of magnitude.

References

1 M. A. Uman, *The Art and Science of Lightning Protection*. New York: Cambridge University Press, 2008.

2 V. Cooray, *Lightning Protection*. IET, 2010.

3 V. A. Rakov and M. A. Uman, *Lightning: Physics and Effects*. Cambridge University Press, 2003.

4 E. P. Krider, R. C. Noggle, and M. A. Uman, "A gated, wideband magnetic direction finder for lightning return strokes," *Journal of Applied Meteorology*, vol. 15, pp. 301–306, 1976.

5 E. A. Lewis, R. B. Harvey, and J. E. Rasmussen, "Hyperbolic direction finding with sferics of Transatlantic origin," *Journal of Geophysical Research*, vol. 65, pp. 1879–1905, 1960.

6 G. N. Oetzel and E. T. Pierce, "VHF technique for locating lightning," *Radio Science*, vol. 4, pp. 199–202, 1969.

7 A. A. Few, "Lightning channel reconstruction from thunder measurements," *Journal of Geophysical Research*, vol. 75, pp. 7517–7523, 1970.

8 M. Rubinstein, C. Romero, F. Rachidi, A. Rubinstein, and F. Vega, "A two-station lightning location method based on a combination of difference of time of arrival and amplitude attenuation," in *2010 Asia-Pacific International Symposium on Electromagnetic Compatibility*, pp. 1154–1157, 2010.

9 A. Mejia, "Lightning location by seismic waves," in *22nd International Conference on Lightning Protection (ICLP)*, Budapest, Hungary, 1994.

10 K. L. Cummins and M. J. Murphy, "An overview of lightning locating systems: history, techniques, and data uses, with an in-depth look at the US NLDN," *IEEE Transactions on Electromagnetic Compatibility*, vol. 51, pp. 499–518, 2009.

11 V. Cooray, *The Lightning Flash*, 2nd edition. IET, 2014.

12 K. Berger, R. B. Anderson, and H. Kroninger, "Parameters of lightning flashes," *Electra*, vol. 80, pp. 23–37, 1975.

13 F. Horner, "Very low frequency propagation and direction finding," *Proceedings of the IEE, Part B*, vol. 104, pp. 73–80, 1957.

14 F. Horner, "The design and use of instruments for counting local lightning flashes," *Proceedings of the IEE., Part B*, vol. 107, pp. 321–330, 1960.

15 N. Mora, F. Rachidi, and M. Rubinstein, "Application of the time reversal of electromagnetic fields to locate lightning discharges," *Atmospheric Research*, vol. 117, pp. 78–85, 2012.

16 G. Lugrin, N. M. Parra, F. Rachidi, M. Rubinstein, and G. Diendorfer, "On the location of lightning discharges using time reversal of electromagnetic fields," *IEEE Transactions on Electromagnetic Compatibility*, vol. 56, pp. 149158, 2014.

17 F. Rachidi and R. Thottappillil, "Determination of lightning currents from far electromagnetic fields," *Journal of Geophysical Research*, vol. 98, pp. 18315–18320, 1993.

18 F. Rachidi, J. L. Bermudez, M. Rubinstein, and V. A. Rakov, "On the estimation of lightning peak currents from measured fields using lightning location systems," *Journal of Electrostatics*, vol. 60, pp. 121–129, 2004.

19 C. A. Nucci, G. Diendorfer, M. Uman, F. Rachidi, M. Ianoz, and C. Mazzetti, "Lightning return stroke current models with specified channel-base current: a review and comparison," *Journal of Geophysical Research*, vol. 95, pp. 20395–20408, 1990.

20 V. A. Rakov and M. A. Uman, "Review and evaluation of lightning return stroke models including some aspects of their application," *IEEE Transactions on Electromagnetic Compatibility*, vol. 40, pp. 403–426, 1998.

21 M. A. Uman and D. K. McLain, "Magnetic field of lightning return stroke," *Journal of Geophysical Research*, vol. 74, pp. 6899–6910, 1969.

22 F. Heidler, "Analytic lightning current functions for LEMP calculations," in *18th International Conference on Lightning Protection ICLP*, Berlin, 1985.

23 V. Cooray, *The Lightning Flash*. IEE, 2003.

24 J. Lojou, N. Honma, K. L. Cummins, R. K. Said, and N. Hembury, "Latest developments in global and total lightning detection," in *2011 7th Asia-Pacific International Conference on Lightning (APL)*, pp. 924–932, 2011.

25 K. Schmidt, H. D. Betz, W. P. Oettinger, M. Wirz, O. Pinto Jr., K. P. Naccarato, *et al.*, "A comparative analysis of lightning data during the EU–Brazil TROCCINOX/TroCCiBras campaign," in *VIII International Symposium on Lightning Protection (SIPDA)*, Sao Paulo, Brazil, 2005.

26 G. Diendorfer, W. Schulz, and F. Fuchs, "Comparison of correlated data from the Austrian lightning location system and measured lightning currents at the Peissenberg Tower," in *24th International Conference on Lightning Protection (ICLP)*, Staffordshire, UK, 1998.

27 H. Karami, F. Rachidi, and M. Rubinstein, "On practical implementation of electromagnetic time reversal to locate lightning," in *23rd International Lightning Detection Conference (ILDC)*, Tucson, Arizona, 2014.

28 T. K. Sarkar and O. Pereira, "Using the matrix pencil method to estimate the parameters of a sum of complex exponentials," *IEEE Antennas and Propagation Magazine*, vol. 37, pp. 48–55, 1995.

7

Electromagnetic Time Reversal Applied to Fault Location in Power Networks

R. Razzaghi, G. Lugrin, M. Paolone, and F. Rachidi

Swiss Federal Institute of Technology (EPFL), Lausanne, Switzerland

7.1 Chapter Overview

The fault location functionality is an important online process required by power systems operation. In transmission networks, this functionality is needed for the identification of the line experiencing a short circuit and to support the consequent reconfiguration of the network to anticipate severe cascading consequences. In distribution networks, fault location is more associated with the quality of service in terms of duration of interruptions when permanent faults occur. With reference to complex power network topologies, such as distribution networks in the presence of an embedded generation, series-compensated high-voltage transmission lines, or multiterminal HVDC networks, existing methods for detecting and locating faults are either inapplicable or they require multiple sensors to achieve acceptable location accuracies. In this chapter, the application of electromagnetic time reversal to fault location in power networks is presented.

Electromagnetic Time Reversal: Application to Electromagnetic Compatibility and Power Systems, First Edition. Edited by Farhad Rachidi, Marcos Rubinstein and Mario Paolone.
© 2017 John Wiley & Sons, Ltd. Published 2017 by John Wiley & Sons, Ltd.
Companion Website: www.wiley.com/go/rachidi55

7.2 Introduction

Power transmission and distribution networks are always prone to short-circuit due to natural events such as falling trees, wind, lightning, ice, storm, or mechanical failure of insulators or other equipment as well as ageing of the insulation systems [1]. Therefore, accurate and reliable fault location processes are required.

In transmission networks, fault location functionality is needed for the identification of the faulty line and the adequate reconfiguration of the network to prevent severe cascading consequences. In distribution networks, fault location is more associated with the quality of service in terms of duration of interruptions when permanent faults occur. Still with reference to distribution networks, the increasing use of distributed generation (DG) units calls for accurate and fast fault location procedures aimed at minimizing the network service restoration time and, consequently, minimizing the unsupplied power.

To comply with the restrictive requirements of modern power systems and to ensure the reliability of the power supply, power systems are equipped with accurate and fast protection and control systems exploiting several fundamental functionalities, including fault location. Since in most cases the failure of the power transfer is due to permanent damages of the system insulation, the first step to repair the it and recover the network is to identify the fault location with the highest possible accuracy. Fast and accurate fault location procedures result in faster repair and recovery of the power supply. The fault location function can be implemented into either [1]: (i) protection relays, (ii) digital fault recorders (DFR), (iii) stand-alone fault locators, or (iv) offline post-fault analysis programs.

It is worth observing that, in general, the time constraint to locate the fault in power systems is less severe than the one required to protection systems since precise fault location requires, in general, complex processing that might take more time compared to protection relays.

The fault location problem has been extensively studied in the literature and numerous methods have been developed for this purpose. At first, this subject has been studied for transmission networks due to the impact of this function on power system operation and the difficulty of locating faults in meshed networks. Then, these studies were extended to distribution networks, due to the increased demand in power quality and reliability. Despite the vast amount of literature, the problem of fault location still presents essential challenges for both transmission and distribution networks:

Transmission networks: the rapid growth in size and complexity of the transmission systems requires accurate fault location methods. In particular, due to the complex and meshed configuration of transmission networks and the impact of the fault location problem on the security of the power systems, fault location is an important and challenging function.

Distribution systems: the fault location problem is even more challenging due to the following reasons (e.g., [1] and [2]):

- Distribution lines are often characterized by the presence of DG units, which can largely affect the accuracy of fault location procedures developed for passive distribution networks.
- Lines are often characterized by a significant geometrical dissymmetry between the phases. Therefore, methods relying on symmetrical components decomposition might provide inaccurate results.
- The effect of the fault impedance on the fault location accuracy can be significant.
- Loads can be unbalanced and the load current, which is superimposed to the fault current, can change during the fault.
- The measurement errors associated with the measurement transformers can be important especially if voltage measurements are used.

Therefore, the fault location problem is still an important ongoing topic of research. In the next section, a brief review of the classical fault location methods is presented.

7.3 Summary of Existing Fault Location Methods

The fault location problem in transmission lines has been a topic of investigation since the 1950s [3] and numerous fault location methods have been proposed in the literature. The various proposed fault location procedures can be classified into three main categories [1, 2]:

1) Methods that analyze pre-fault and post-fault voltage/current phasors (phasor-based methods).
2) Methods that analyze fault-originated electromagnetic transients of currents and/or voltages (traveling wave-based methods).
3) Knowledge-based approaches (artificial intelligence methods).

7.3.1 Impedance-Based Methods

These methods rely on the calculated impedance of the faulty line to identify the fault location by measuring the voltage and current phasors. The methods belonging to this category can be further classified into the following subcategories: (i) single-terminal measurement methods (e.g., [4] to [7]), (ii) two-terminal measurement methods (e.g., [8] to [12]), and (iii) multiterminal algorithms (e.g., [13] to [17]) that employ measurements from multiple ends of multiterminal transmission lines. Two-terminal and multiterminal measurement-based fault location methods can be based on either unsynchronized (e.g., [10, 18], and [19]) or synchronized (e.g., [8] and [16]) voltage/current measurements.

Single-terminal measurement-based fault location methods estimate the fault distance by using voltage and current measurements at a particular end of the line. These approaches are simpler compared to two-terminal or multiterminal measurement-based approaches since they do not require communication means. Therefore, they are more attractive for practical applications. However, the solution of the fault location problem requires several assumptions and simplifications [4, 5], which impact the fault location accuracy [6]. The accuracy of these methods depends on the fault resistance, load, configuration of the line, and load flow unbalance.

On the other hand, the methods based on two-/multiterminal measurements provide more accurate and robust fault location results compared to single-terminal methods. However, they require communication links to exchange the data between multiple ends. The availability of multiple data enables the effect of the fault resistance uncertainty and other affecting parameters to be minimized, and therefore to minimize the fault location estimation error. Nevertheless, the performance of these methods is mainly dependent on efficient communication links which add non-negligible complexity to the system. Multiterminal fault location methods require a global positioning system (GPS) to provide a common time reference. Their performance can also be affected by the accuracy and the loss of the GPS signal [12].

Despite the straightforward solutions provided by the impedance-based fault location methods, their accuracy might be affected by the fault resistance, configuration of the line, load flow unbalance, and the presence of DG units [2]. Indeed, active distribution networks (ADNs) in which DG units are connected to the distribution system feeders can largely affect the accuracy of these procedures. Moreover, these methods are not applicable to HVDC transmission systems, which are one of the essential elements of modern power systems.

7.3.2 Traveling Wave-Based Methods

Traveling wave-based methods have been increasingly investigated in the literature (e.g., [20] to [25]). These methods rely on the analysis of the high-frequency components of the fault-originated transient signals, which are little influenced by the fault impedance [26]. In particular, these methods are considered to be the most appropriate to identify fault location in DC transmission systems (e.g., [27]).

Traveling wave-based fault location methods analyze different features of the traveling waves and utilize various techniques to identify the fault location. One of the first adopted techniques is based on the cross-correlation between the forward and backward traveling waves [21]. The main drawback of such a method is the discrimination of the traveling waves originating from the fault point from reflections at remote ends of the line [28]. Moreover, the accuracy of this method is mainly dependent on the sampling window [28]. An improved correlation-based method was proposed in [28] which uses combined short and long window lengths. The arrival time-based methods identify the fault location by assessing the arrival time of the traveling waves at one (single-terminal) or different terminals of the line (multiterminal) [29].

In single-terminal methods, the arrival times of the initial and reflected traveling waves at one single terminal of the line are used. The fault location is identified by assessing the time delay between successive reflections of the measured traveling wave signals (e.g., [29] and [30]). These methods avoid the cost and complexities associated with multiterminal measurement synchronization and communication links. The main difficulty related to single-terminal methods is the detection and discrimination of the fault-originated waves from the reflections associated with other terminals [29]. Moreover, these methods commonly have problems for locating faults located very close to the observation point [31].

Multiterminal methods, which rely on multiple measurement locations, provide better fault location accuracy compared to one-terminal methods. For the case of a single line, if the measurements at the two terminals are synchronized, the difference between the arrival times of the traveling waves at the two terminals can be used to identify the fault location. To this end, a precise GPS signal is required to synchronize the measurements at the two terminals and the wave velocity should be known [2, 32]. Nevertheless, despite several advantages provided by two-terminal synchronized fault location methods (e.g., insensitivity to the variations of source impedances, fault distances, and fault impedances), practically speaking, their applications are limited due to the lack of a common time reference in all substations [33]. To cope with this issue, unsynchronized measurement methods for the fault location problem have been proposed (e.g., [31] and [33]).

Wavelet analysis is a powerful signal processing tool able to analyze the signal in both time and frequency domains. A particular feature of the wavelet transformation (WT) is the automatic altering of the data window according to the frequency. WT has been successfully applied to overcome the shortcomings associated with the traveling wave-based methods (e.g., [24, 26], and [34]). First, discrete-wavelet transform (DWT) was used to identify the fault location due to its straightforward implementation and reduced computation time [35]. However, compared to DWT, the continuous-wavelet transform (CWT) provides more detailed and continuous analysis of the signal. This is the result of smooth shifting of the analyzed wavelet over the full domain of the signal compared to the dyadic shift in

DWT [36]. Therefore, CWT-based fault location methods provide more precise results (e.g., [26, 34], and [35]). The limitation associated with these methods is related to the use of traditional mother wavelets, which does not allow identifying all the characteristic frequencies of the traveling waves. In [26], a method has been proposed to overcome this problem by building specific mother wavelets inferred from the fault-originated transient waveforms. Nevertheless, the application of WT-based fault location methods requires a considerable amount of computational effort.

Despite the superior performance of the traveling wave-based methods compared to phasor-based methods, their accuracy might still be affected by the following factors [2]:

- Assessment of the number of observation points versus the number of possible multiple fault location solutions.
- Requirement of a precise time stamping for methods requiring multiple synchronized metering stations.
- Loss of GPS signal impacting the fault location accuracy.
- Requirement of large bandwidth measurement systems.

7.3.3 Knowledge-Based Methods

In addition to the previous approaches, research efforts have been devoted to the use of knowledge-based fault location methods. Expert systems identify the most probable fault location by means of available information regarding the network status (e.g., the state of switches, unpowered user complaints, etc.) (e.g., [37] and [38]). Artificial neural networks (ANN) and fuzzy logic have been widely studied for the fault location problem (e.g., [39] and [40]). Nevertheless, the extensive training of such methods limits their application to real systems.

7.4 Application of Electromagnetic Time Reversal (EMTR) for the Fault Location Problem

7.4.1 Basic Concepts and Time-Reversal Invariance of Telegrapher's Equations

To overcome the limitations associated with the existing traveling wave-based fault location methods (e.g., requirement of time-synchronized, multiterminal measurement stations and complexity due to the sophisticated signal processing techniques), time reversal was recently considered as a method to locate faults in power networks.

In order to apply the time-reversal process to the fault location problem, first we need to examine the properties of the transmission line wave equations under the time-reversal operator. The voltage wave equation for a multiconductor, lossless transmission line reads as [41, 42]

$$\frac{\partial^2}{\partial x^2} \mathbf{U}(x, t) - \mathbf{L'C'} \frac{\partial^2}{\partial t^2} \mathbf{U}(x, t) = 0 \qquad (7.1)$$

where $\mathbf{U}(x, t)$ is a vector containing the phase-to-ground voltages at position x and time t and $\mathbf{L'}$ and $\mathbf{C'}$ are the matrices of per-unit-length inductance and capacitance of the line, respectively.

Time reversing the wave equation yields

$$\frac{\partial^2}{\partial x^2} \mathbf{U}(x, -t) - \mathbf{L'C'} \frac{\partial^2}{\partial t^2} \mathbf{U}(x, -t) = 0 \qquad (7.2)$$

Therefore, if $\mathbf{U}(x, t)$ is a solution of the wave equation, then $\mathbf{U}(x, t)$ is also a solution. In other words, the wave equation is invariant under a time-reversal transformation if there is no energy absorption during propagation in the medium. In our specific application, this hypothesis is satisfied if the transmission line is lossless. The inclusion of losses in the transient propagation with respect to the EMTR applicability to locate faults is further discussed in Section 7.5.

In practical implementations, a signal $s(x, t)$ is necessarily measured only during a finite period of time from an initial time selected here as the origin $t = 0$ to a final time $t = T$, where T is the duration of the signal. To enforce the argument of the time-reversed variables to be positive for the duration of the signal, we will consider, in addition to time reversal, an additional time delay T:

$$s(x, t) \mapsto s(x, T - t) \qquad (7.3)$$

In order to illustrate the EMTR application to the fault location problem, a brief explanation of the electromagnetic transients associated with faults in power systems is given here.

A fault event in a power system can be associated with an injection into the power system itself of a step-like wave initiated by the fault occurrence. The fault-generated waves travel along the lines of the network and get reflected at the line extremities, which are characterized by reflection coefficients whose values depend on the line surge impedance (characteristic impedance) and the input impedances of the power components connected at the line extremities. In particular, the line extremities can be grouped into three categories, namely line terminals with power transformers, junctions to other lines, and the fault location. As discussed in [34] and [43], for each of these boundary conditions the following considerations can be made:

- Extremities, where a power transformer is connected, can be assumed, for the traveling waves, to be open circuits, and therefore the relevant voltage reflection coefficient is close to +1; indeed, fault-originated traveling waves are characterized by a spectrum with high-frequency components for which the input impedance of power transformers is generally dominated by a capacitive behavior with capacitance values on the order of a few hundreds of pF (e.g. [43]).
- Extremities that correspond to a junction between more than two lines are characterized by a negative reflection coefficient.
- The reflection coefficient of the extremity where the fault is occurring is close to −1, as the fault impedance can be assumed to be significantly smaller than the line surge impedance.

With the above considerations and for a given network topology, it is possible to determine a certain number of paths p, each one delimited between two extremities. Figure 7.1 illustrates these paths for a simplified network topology composed of a main feeder and a lateral. A given observation point in the system where voltage or current waveforms are measured will observe a superposition of traveling waves associated with the various paths.

Therefore, one can conclude that the domain of application of any fault location method belonging to the traveling wave-based methods described in Section 7.3.2 is formed by a one-dimensional space (associated with the line longitudinal coordinate x) with given boundary conditions. In what follows, the proposed fault location method based on EMTR is presented.

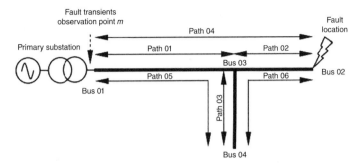

Figure 7.1 Paths covered by traveling waves caused by a fault at Bus 02. (Adapted from [44].)

7.4.2 EMTR-Based Fault Location Method

The application of the EMTR to locate faults in a power network will be based on the following three steps: (i) measurement of the fault-originated electromagnetic transient in a single observation point, (ii) simulation of the back-injection of the time-reversed measured fault signal for different guessed fault locations (GFLs) using the network model, and (iii) assessment of the fault location by determining, in the network model, the point characterized by the largest energy concentration associated with the back-injected time-reversed fault transients. In what follows, we illustrate the analytical aspects related to the proposed EMTR-based fault location method.

As described in [45] and [46], one of the main hypotheses of the time-reversal (TR) method is that the topology of the system needs to remain unchanged during the transient phenomenon of interest. Fault transients in power networks do not satisfy such a condition as the presence of the fault itself involves a change in the network topology when the fault occurs (i.e. at $t = t_f$). However, for reversed times t such that $t < T - t_f$, EMTR is still applicable if the guessed fault is considered at the correct location. On the other hand, for a guessed location that does not coincide with the real one, time-reversal invariance does not hold. As a result of this property, time-reversed back-propagated signals will combine constructively to reach a maximum at the correct

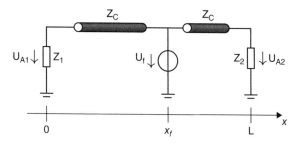

Figure 7.2 Simplified representation of the post-fault line configuration for the EMTR analytical validation.

fault location. This property will be validated in the next sections to analytically prove the method and is used in both experimental measurements and simulation test cases.

7.4.2.1 Frequency-Domain Derivation

The aim of this subsection is to describe analytically the behavior of the line response after a fault. In order to express analytically the line response, the problem is formulated in the frequency domain. In particular, to provide a more straightforward description of the EMTR technique, we will make reference to a single-conductor lossless transmission line (Figure 7.2) of length L. The line parameters may refer to a typical overhead transmission line. In particular, the surge impedance (characteristic impedance) is on the order of a few hundred ohms. We assume that at both line extremities power transformers are connected.

Therefore, as discussed before, they are represented by means of high input impedances (Z_1 and Z_2 in Figure 7.2). The fault coordinate is x_f and fault transient waveforms are assumed to be recorded either at one end or at the two ends of the line. As the line model is lossless, the damping of the transients is provided only by the fault impedance, if any, and the high terminal impedances Z_1 and Z_2.

As the analyzed fault transients last for only a few milliseconds, we assume that the pre-fault condition of the line is characterized by a constant value of voltage all along the line length ($0 \leq x \leq L$).

To specify the boundary conditions of the two line sections of Figure 7.2, namely for $0 \leq x \leq x_f$ and $x_f \leq x \leq L$, we can define reflection coefficients at $x = 0$ ($i = 1$ of Figure 7.2) and $x = L$ ($i = 2$ of Figure 7.2) as

$$\rho_i = \frac{Z_i - Z_C}{Z_i + Z_C}; \quad i = 1, 2 \tag{7.4}$$

Without losing generality, coefficients ρ_i in (7.4) could be assumed to be frequency-independent within the considered short observation time. Concerning the boundary condition at the fault location, we assume to represent it by means of a voltage source $U_f(\omega)$ located at $x = x_f$. Here, for the sake of abstraction, we represent the fault by means of an ideal voltage source with zero internal impedance that, as a consequence, represents a solid fault. Therefore, the voltage reflection coefficient at this point of the line is $\rho_i = -1$. Additionally, in view of the lossless line assumption, the line propagation constant, γ, is purely imaginary, namely: $\gamma = j\beta$, with $\beta = \omega/c$ (c being the speed of light). The analytical expressions of the voltages observed at the line terminals $x = 0$ and $x = L$ in the frequency domain read as

$$U_{A1}(\omega) = U(0, \omega) = \frac{(1 + \rho_1)e^{-\gamma x_f}}{1 + \rho_1 e^{-2\gamma x_f}} U_f(\omega) \tag{7.5}$$

$$U_{A2}(\omega) = U(L, \omega) = \frac{(1 + \rho_2)e^{-\gamma(L - x_f)}}{1 + \rho_2 e^{-2\gamma(L - x_f)}} U_f(\omega) \tag{7.6}$$

Note that the effect of the wire and ground losses can be represented as an additional frequency-dependent longitudinal impedance [47]. However, except for the case of distributed exciting sources (such as those produced by a nearby lightning discharge), ground and wire losses can be disregarded for typical overhead power lines [48].

Now the EMTR process can be carried out by using either (i) two observation points located at both extremities of the line or (ii) one observation point at one end of the line.

7.4.2.1.1 Frequency-Domain Application of EMTR by Considering Two Observation Points at Each Line Terminal

According to the concept of a time-reversal mirror (TRM) [49], a number of observation points at which transient signals

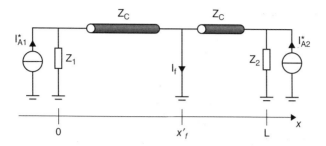

Figure 7.3 Representation of the EMTR applied to the single-line model of Figure 7.2.

initiated by the source (here: the fault) are measured could be used to apply the TR process. In a first step, it is assumed that two observation points at both ends of the line are used.

Equations (7.5) and (7.6) provide the frequency-domain expressions of fault-originated voltages at two observation points located at the line terminals. In agreement with the EMTR method, we can replace these observation points with two sources, each one imposing the time-reversed voltage fault transients. As shown in Figure 7.3, the TR operator in the frequency domain is represented by the complex conjugate of the Fourier transform of the signal. Therefore, the time-reversed recorded voltage transients are $U_{A1}^*(\omega)$ and $U_{A2}^*(\omega)$, where $*$ denotes the complex conjugate operator. Here we consider the Norton equivalents as

$$I_{A1}^* = \frac{U_{A1}^*(\omega)}{Z_1} \tag{7.7}$$

$$I_{A2}^* = \frac{U_{A2}^*(\omega)}{Z_2} \tag{7.8}$$

where I_{A1}^* and I_{A2}^* are the injected currents as shown in Figure 7.3. As the location of the fault is the unknown of the problem, we will place it at a generic location x_f'. The contributions in terms of currents at the unknown fault location x_f' coming from the first and the second time-reversed sources I_{A1}^* and I_{A2}^* are given respectively by

$$I_{f1}(x_f', \omega) = \frac{(1 + \rho_1)e^{-\gamma x_f'}}{1 + \rho_1 e^{-2\gamma x_f'}} I_{A1}^*(\omega) \tag{7.9}$$

$$I_{f2}(x_f', \omega) = \frac{(1 + \rho_2)e^{-\gamma(L - x_f')}}{1 + \rho_2 e^{-2\gamma(L - x_f')}} I_{A2}^*(\omega) \tag{7.10}$$

Introducing (7.5) to (7.8) into (7.9) and (7.10), we obtain

$$I_{f1}(x_f', \omega) = \frac{(1 + \rho_1)^2 \, e^{-\gamma(x_f' - x_f)}}{Z_1 \left(1 + \rho_1 e^{-2\gamma x_f'}\right)\left(1 + \rho_1 e^{+2\gamma x_f}\right)} U_f^*(\omega) \tag{7.11}$$

$$I_{f2}(x_f', \omega) = \frac{(1 + \rho_2)^2 \, e^{-\gamma(x_f' - x_f)}}{Z_2 \left(1 + \rho_2 e^{-2\gamma(L - x_f')}\right)\left(1 + \rho_2 e^{+2\gamma(L - x_f)}\right)} U_f^*(\omega) \tag{7.12}$$

Therefore, we can derive a closed-form expression for the total current flowing through the guessed fault location (GFL) x_f':

$$I_f(x_f', \omega) = I_{f1}(x_f', \omega) + I_{f2}(x_f', \omega) \tag{7.13}$$

In what follows, we will make use of (7.13) to show the capability of the EMTR to converge the time-reversed injected transients to the fault location.

Let us make reference to a line characterized by a total length $L = 10$ km and let us assume a fault occurring at $x_f = 8$ km. The line is characterized by terminal impedances $Z_1 = Z_2 = 100$ kΩ and, for the fault, we assume $U_f = 1/j\omega$ V/(rad/s). The line is lossless and the per-unit-length capacitance and inductance are $C = 7.10 \times 10^{-12}$ F/m and $L = 1.56 \times 10^{-6}$ H/m, respectively.

By moving x_f' from 0 to L, it is possible to compute the current at the GFLs using (7.13). Figure 7.4 shows the normalized fault current signal energy (FCSE) (where the normalization has been implemented with respect to the maximum signal energy value of I_f for all the guessed fault locations) within a frequency-spectrum ranging from DC to 1 MHz. From Figure 7.4, it is clear

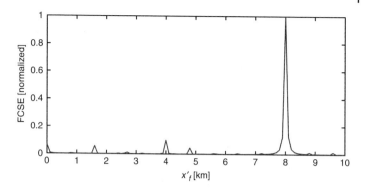

Figure 7.4 Normalized fault current signal energy (FCSE) as a function of the guessed fault location (GFL) x'_f with multiple observation points. The real fault location is at $x'_f = 8$ km. (Adapted from [41].)

that the energy of $I_f(x'_f, \omega)$ reaches its maximum when the GFL coincides with the real fault location.

7.4.2.1.2 Frequency-Domain Application of EMTR by Considering One Observation Point

As was mentioned before, one of the main problems in power systems protection is the limited number of observation points where measuring equipment can be placed. Therefore, the demonstration that the EMTR-based fault location method could be applied also for the case of a single observation point is of importance. To this end, let us assume that the fault-originated electromagnetic transients are observed only at one location, namely at the line left terminal. The network schematic in the time-reversal state will be the one in Figure 7.5.

By making reference to the configuration of the previous case, we can extend the procedure to the case where only one injecting current source (I_{A1}) is considered. In particular, we can derive from (7.14) the fault current at any guessed fault location x'_f as follows:

$$I_f(x'_f, \omega) = I_{f1}(x'_f, \omega) = \frac{(1 + \rho_1)^2 \, e^{-\gamma(x'_f - x_f)}}{Z_1 \left(1 + \rho_1 e^{-2\gamma x'_f}\right) \left(1 + \rho_1 e^{+2\gamma x_f}\right)}$$
$$\times \, U_f^*(\omega) \tag{7.14}$$

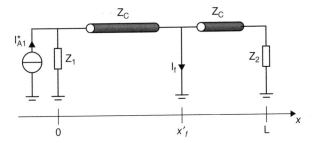

Figure 7.5 Representation of the EMTR applied to the single-line model of Figure 7.2 where a single observation point is placed at the beginning of the line ($x = 0$).

Figure 7.6 shows the normalized FCSE of I_f (where the normalization has been implemented with respect to the maximum signal energy value of I_f for all the GFLs) within a frequency-spectrum ranging from DC to 1 MHz.

It can be noted that the energy of $I_f(x'_f, \omega)$ is maximum at the fault location even for the case of a single observation point. From Figure 7.4 and Figure 7.6, it can further be observed that the two curves, corresponding respectively to two and to one observation points, provide the correct fault location accurately

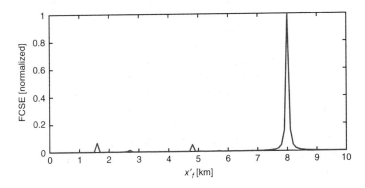

Figure 7.6 Normalized FCSE as a function of the GFL x'_f with a single observation point. The real fault location is at $x'_f = 8$ km. (Adapted from [41].) An animation version of this figure can be found at www.wiley.com/go/rachidi55.

and the method can be effectively applied using a single observation point.

An animation can be found here showing the back-injected voltage signal from the observation point and the evolution of the fault current and FCSE in different GFLs.

7.4.2.2 Time-Domain Algorithm

In the previous section, we have derived closed-form expressions for the fault current as a function of the guessed fault location. The purpose of this section is to extend the proposed method to realistic time-domain cases. The flowchart shown in Figure 7.7

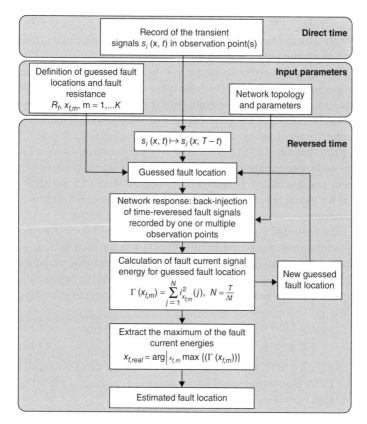

Figure 7.7 Flowchart of the proposed EMTR-based fault location method. (Adapted from [41].)

illustrates the step-by-step fault location procedure based on EMTR.

It can be seen that the proposed procedure, similarly to other methods proposed in the literature (e.g., [24]), requires knowledge of the network topology as well as its line parameters. Such knowledge is used to build a corresponding network model.

We assume to record fault transients, $s_i(x, t)$ (with $i = 1, 2, 3$ for a three-phase system), at a generic observation point located inside the part of the network with the same voltage level comprised between transformers. The transient signals initiated by the fault is assumed to be recorded within a specific time window:

$$s_i(t), \quad t \in [t_f, t_f + T] \tag{7.15}$$

where t_f is the fault triggering time and T is the recording time window large enough to damp-out $s_i(t)$.

The unknowns of the problem are the fault type, location, and impedance. Concerning the fault type, we assume that the fault location procedure will operate after the relay maneuver. Therefore, the nature of the fault (single- or multiphase) is assumed to be known. Concerning the fault location, we assume a set of a priori locations (GFLs) $x_{f,m}, m = 1, \dots, K$ for which the EMTR procedure is applied. For all the GFLs, an a priori value of the fault resistance, R_{xf}, is assumed. As will be shown in the application examples and performance evaluation section, the fault location accuracy is not sensitive to the fault impedance.

The recorded signals are reversed in time and back-injected from the observation point into the system for each GFL $x_{f,m}$. In order to make the argument that the time-reversed variables be positive for the duration of the signal, we add, in addition to time reversal, a time delay equal to the duration of the recording time T:

$$\hat{t} = (T + t_f) - t \tag{7.16}$$

$$\bar{s}(\hat{t}), \hat{t} \in [0, T] \tag{7.17}$$

As shown in Figure 7.7, for each of the GFLs, we can compute the FCSE that corresponds to the energy of the currents flowing through the GFL as

$$\Gamma(x_{f,m}) = \sum_{p=1}^{M} \sum_{j=1}^{N} \left[i_{x_{f,m}}^p (j) \right]^2, \quad T = N\Delta t \tag{7.18}$$

where N is the number of samples and Δt the sampling time, and p indicates the number of conductors in the line that are involved in the fault. According to the EMTR method presented in the previous section, the energy given by (7.18) is maximized at the real fault location. Thus, the maximum of the calculated FCSEs will indicate the real fault point:

$$x_{f,real} = \arg|_{x_{f,m}} \max\{(\Gamma(x_{f,m}))\} \tag{7.19}$$

7.5 The Issue of Losses: Back-Propagation Models

As shown by Equation (7.2), Telegrapher's equations are invariant under a time-reversal transformation for lossless lines. Electromagnetic propagation involving a dissipative medium is not rigorously time-reversal invariant unless an inverted-loss medium is considered for the reverse time. In this section, the effect of the line losses on the accuracy of the fault location method is analyzed by considering three different back-propagation models: lossless, lossy, and inverted-loss [50]. For the sake of simplicity, we make reference to Figure 7.8, which represents the equivalent circuit of a differential length of a single-wire line above a ground plane [51].

In Figure 7.8, L', C', and G' are the per-unit-length longitudinal inductance, transverse capacitance, and transverse conductance, respectively, Z'_w is the per-unit-length internal impedance

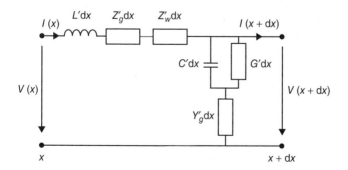

Figure 7.8 Equivalent circuit of a single-wire line above a ground plane.

of the wire, and Z'_g and Y'_g are the per-unit-length ground impedance and admittance, which account respectively for the losses associated with the penetration of magnetic and electric fields in the ground [51].

The effect of losses (in the conductor and in the ground) results essentially in an attenuation of propagated transients and a modification of the propagation speed, both effects being generally frequency dependent. In the application of the EMTR to fault location in which the timing is crucial, the modification of the propagation speed is expected to be more critical than the attenuation of the amplitude [50]. In what follows, three different back-propagation models will be investigated [50].

7.5.1 Inverted-Loss Back-Propagation

This model is equivalent to invert the real part of the propagation constant:

$$\tilde{\gamma} = -\alpha + j\beta \tag{7.20}$$

Note that this is not a physical model since the line itself becomes active and gives energy to the signal that is propagating along it. However, this model can be numerically implemented. In this case, it can readily be shown that Telegrapher's equations are time-reversal invariant. An exact location is hence expected as a result of the application of this model, under the assumption that the line parameters are perfectly known.

7.5.2 Lossless Back-Propagation

In this model, the losses in the back-propagation are disregarded. In other words, the line per-unit-length parameters during back-propagation become

$$\tilde{Z}'_w = 0, \tilde{Z}'_g = 0, \tilde{G}' = 0, \tilde{Y}'_g \to \infty \tag{7.21}$$

Hence, the propagation and phase velocity will in general not be the same as during the direct propagation, leading possibly to inaccuracies in the determination of the fault location.

The resulting propagation constant for the back-propagation is therefore

$$\tilde{\gamma} = j\beta \tag{7.22}$$

7.5.3 Lossy Back-Propagation

In this model, a lossy model for the line is used for the back-propagation. The back-propagation line per-unit-length parameters become in this case

$$\tilde{Z}'_w = Z'_w, \quad \tilde{Z}'_g = Z'_g, \quad \tilde{G}' = G', \quad \tilde{Y}'_g = Y'_g \tag{7.23}$$

and the resulting propagation constant for the back-propagation is

$$\tilde{\gamma} = \gamma = \alpha + j\beta \tag{7.24}$$

Even though a lossy medium is not time-reversal invariant, the propagation speed, which is a key parameter, remains unchanged. Therefore, it can be anticipated that all contributions from discontinuities that would occur in a real network will add up in phase at the fault location. For this reason, we can expect more accurate results with this model than those associated with a lossless back-propagation model.

7.5.4 Comparison of the Back-Propagation Models

To compare the performance of the three models described in the previous section, we will consider a simple configuration of a 10-km long, single-wire overhead line above a finitely conducting ground. The assumed fault is located at 8 km from the left terminal where the voltage transient generated by the fault is recorded. The propagation in direct time takes the losses into account, and the back-propagation is simulated making use of each of the three models presented previously. The computation is made in the frequency domain, in the range 1 kHz–1 MHz and is implemented in Matlab®. The parameters of the line are given in Table 7.1.

The per-unit-length parameters of the line were computed using expressions that can be found in [48] and [51]. The magnitude of the total per-unit-length longitudinal impedance Z' is

Table 7.1 Parameters of the line.

Parameter	Value
Height above the ground	10 m
Diameter of the wire	1 cm
Conductivity of the wire (copper)	5.8×10^7 S/m
Relative permittivity of the ground	10
Conductivity of the ground	10^{-2}–10^{-3} S/m
Terminal resistances	50 kΩ

plotted in Figure 7.9 as a function of the frequency. In the same plot, we have also shown the contributions of the inductive term and ground and wire impedances. It can be seen that the ground losses are dominant compared to the losses in the conductor [47].

The magnitude of the total per-unit-length admittance Y' is plotted in Figure 7.10. It can be seen that the capacitive term is dominant and the losses due to the transverse air and ground admittances are negligible [48].

Figure 7.11 shows the plot of the phase and group velocities as a function of the frequency, when considering losses in the line.

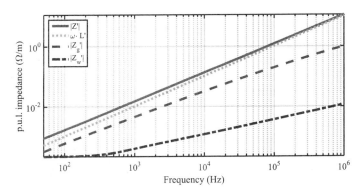

Figure 7.9 Magnitude of the per-unit-length longitudinal impedance as a function of frequency. The ground conductivity is $\sigma_g = 0.001$ S/m. (Adapted from [52].)

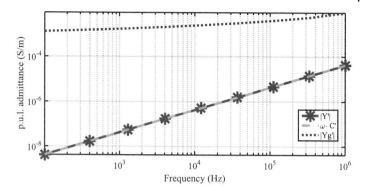

Figure 7.10 Magnitude of the per-unit-length transverse admittance as a function of frequency. Ground conductivity is $\sigma_g = 0.001$ S/m. (Adapted from [52].)

Since the considered line is overhead, the velocity when disregarding losses is equal to the speed of light in vacuum. Now, considering the proposed EMTR-based fault location method for this network, the energy of the current flowing from the conductor to the ground in the back-propagated time at different GFLs along the line are calculated. The simulations are carried out using the three back-propagation models described earlier and considering two different values for the ground conductivity, $\sigma_g = 0.01$ S/m and $\sigma_g = 0.001$ S/m.

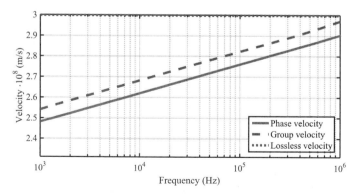

Figure 7.11 Phase and group velocities as a function of the frequency. Ground conductivity is $\sigma_g = 0.001$ S/m. (Adapted from [52].)

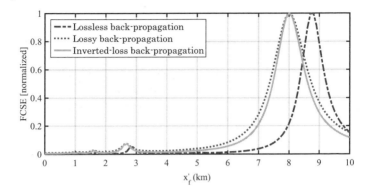

Figure 7.12 Fault current energy normalized to its maximum for the three back-propagation models. Ground conductivity $\sigma_g = 0.01$ S/m. (Adapted from [52].)

Figure 7.12 and Figure 7.13 show the FCSE normalized to its maximum for the three back-propagation models. The ground conductivities are considered at $\sigma_g = 0.01$ and $\sigma_g = 0.001$ S/m, respectively.

It can be seen that the lossy and inverted-loss models were able to locate the fault at the correct position (8 km). On the other hand, the lossless back-propagation model was not able to accurately locate the fault. The location errors for the two

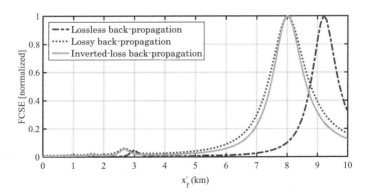

Figure 7.13 Fault current energy normalized to its maximum for the three back-propagation models. Ground conductivity $\sigma_g = 0.001$ S/m. (Adapted from [52].)

Table 7.2 Location error according to the three models of back-propagation.

	Error (km)	
Back-propagation model	$\sigma_g = 0.01$ S/m	$\sigma_g = 0.001$ S/m
Lossless back-propagation	0.8	1.3
Lossy back-propagation	0	0
Inverted-loss back-propagation	0	0

considered values for the ground conductivity (0.01 S/m and 0.001 S/m) were, respectively, 800 m and 1.3 km Table 7.2). By considering this example, we can conclude that, as expected, an inverted-loss model for the back-propagation results in a perfect estimation of the fault location.

It is also observed that a back-propagation model in which the losses are included results also in a perfect estimation of the fault location, even though in this case Telegrapher's equations are not strictly time-reversal invariant. This is a very significant result since it allows the use of commercial codes to simulate the back-propagation phase during which the time-reversed fault-generated transients are injected into the network [50].

7.6 Experimental Validation

To provide a ground truth validation of the EMTR-based fault location method, an experimental test was carried out by making reference to a reduced-scale coaxial cable system [41]. Such a system was realized by using standard RG-58 and RG-59 coaxial cables where real faults were hardware-initiated. The topologies adopted to carry out the experimental validation are shown in Figure 7.14.

As seen in the figure, the first topology corresponds to a single transmission line whilst the second one corresponds to a T-shape network where the various branches are composed of both RG-58 and RG-59 cables (i.e., each branch has a different surge impedance and propagation speed). Figure 7.14 shows also the GFLs at which the current flowing through the fault was

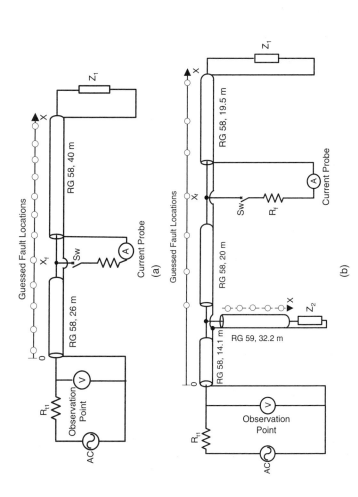

Figure 7.14 Topologies adopted for the reduced-scale experimental setup: (a) a single transmission line configuration (RG-58 coaxial cable), (b) a T-shape network made of both RG-58 and RG-59 coaxial cables. (Adapted from [41].)

measured. For each considered topology, transients generated by the fault were recorded at one observation point, shown also on Figure 7.14.

The fault-originated transients were measured by means of a 12-bit oscilloscope (LeCroy Waverunner HRO 64Z) operating at a sampling frequency of 1 GSa/s (giga samples per second). For the direct time, the oscilloscope directly records voltages at the shown observation points marked in Figure 7.14a and b. For the reversed time, the current at each GFL was measured by using a 2877 Pearson current probe characterized by a transfer impedance of 1 Ω and an overall bandwidth of 300 Hz–200 MHz. It is worth observing that the switching frequencies for the adopted reduced-scale systems are on the order of a few MHz.

The time-reversed transient waveforms were generated by using a 16-bit arbitrary waveform generator (LeCroy ArbStudio 1104) operating at a sampling frequency of 1 GSa/s (the same adopted to record the fault-originated waveforms). The lines were terminated by high impedances (Z_1 and Z_2 equal to 1 MΩ) and the voltage source injecting the time-reversed signal was connected to the line through a lumped resistance of $R = 4.7$ kΩ in order to emulate, to a first approximation, the high-input impedance of power transformers with respect to fault transients.

The faults were generated at an arbitrary point on the cable network. They were realized by a short-circuit between the coaxial cable shield and the inner conductor. It is important to underline that such types of faults excite the shield-to-inner conductor propagation mode that is characterized, for the adopted coaxial cables, by propagation speeds of 65.9% (RG-58) and 82% (RG-59) of the speed of light c. It is worth noting that the limited lengths of the reduced-scale cables (i.e., tens of meters) involve propagation times in the order of tens to hundreds of nanoseconds. Such a peculiarity requires that the fault emulator needs to be able to change its status in a few nanoseconds in order to correctly emulate the fault. The chosen switch was a high-speed MOSFET (TMS2314) with a turn-on time of 3 ns. The MOSFET was driven by a National Instruments digital I/O card C/series 9402 able to provide a gate signal to the MOSFET with a subnanosecond risetime and a maximum voltage of 3.4 V. The schematic representation of the circuit of the hardware fault emulator and the built PCB board is illustrated in Figure 7.15a and b,

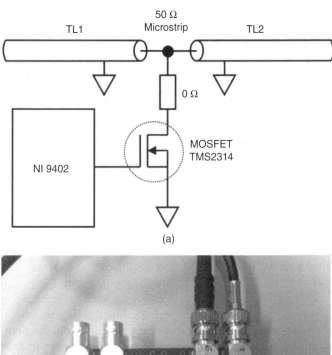

(a)

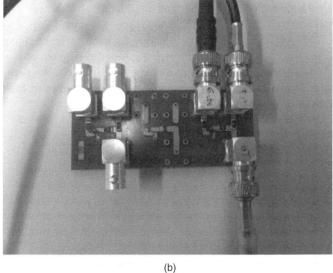

(b)

Figure 7.15 MOSFET-emulated fault adopted in the reduced-scale experimental setup: (a) schematic representation, (b) built PCB board.

Figure 7.16 The experimental setup used for the EMTR-based fault location method validation.

respectively. Note that the experiment reproduces solid faults since no resistors were placed between the MOSFET drain and the transmission line conductors. The experimental setup is depicted in Figure 7.16.

By making reference to the topology of Figure 7.14a, Figure 7.17a shows the measured direct-time voltage at the considered observation point for a fault location $x_f = 26$ m. The measured voltage was then time-reversed and injected back into the line using the arbitrary waveform generator for each of the 12 different guessed fault locations that are indicated in Figure 7.14a. For each case, the fault current resulting from the injection of the time-reversed signal of Figure 7.17a was measured using the Pearson current probe. Figure 7.17b to d show the waveforms of the fault current at the guessed fault locations $x'_f = 23$ m, $x'_f = 26$ m and $x'_f = 28$ m, respectively, resulting from the injection of the time-reversed signal.

The normalized FCSE is shown in Figure 7.18 as a function of the GFL (also in this case, the normalization has been implemented with respect to the maximum signal energy of the fault

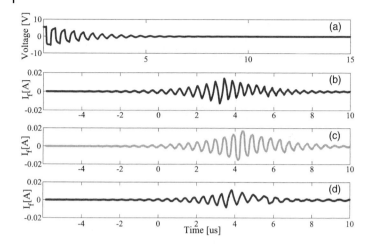

Figure 7.17 Experimentally measured waveforms for a fault location $x_f = 26$ m for the topology of Figure 7.14a: (a) direct-time voltage measured at the observation point located at the beginning of the line. Measured fault currents as a result of the injection of time-reversed signal at guessed fault locations: (b) $x'_f = 23$ m, (c) $x'_f = 26$ m (real fault location), and (d) $x'_f = 28$ m. (Adapted from [41].)

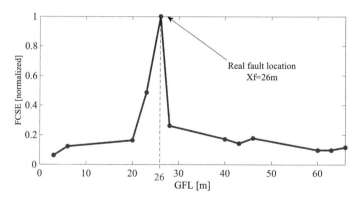

Figure 7.18 Normalized FCSE as a function of the position of the GFL for the configuration shown in Figure 7.14a. The real fault location is at $x_f = 26$ m. (Adapted from [41].)

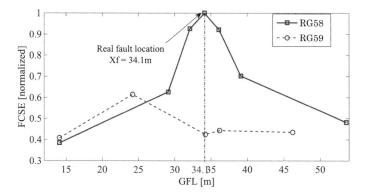

Figure 7.19 Normalized FCSE as a function of the position of the GFL for the case of the topology presented in Figure 7.14b. The real fault location is at $x_f = 34.1$ m in RG-58. (Adapted from [41].)

current in the guessed fault location). It can be observed that the correct fault location is unequivocally identified.

Figure 7.19 shows the same signal energy profiles for the case of the topology of Figure 7.14b. In this case, the real fault location was at a distance of 34.1 m from the source and in the RG-58 section of the network. It can be observed that, as also in the case of a multibranched network with lines characterized by different electrical parameters (i.e. inhomogeneous lines with different surge impedances), the proposed methodology correctly identifies the fault location.

7.7   Case Studies and Performance Evaluation

7.7.1   Inhomogeneous Network Composed of Mixed Overhead Coaxial Cable Lines

For the first application example, reference is made to the case of a network composed of a three-conductor transmission line and an underground coaxial cable, shown in Figure 7.20.

The overhead line length is 9 km and the cable length is 2 km. They are modeled by means of a constant-parameter model implemented within the EMTP-RV simulation environment [53–55]. Both the overhead line and the cable parameters have been inferred from typical geometries of 230 kV lines and

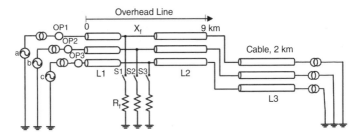

Figure 7.20 Schematic representation of the inhomogeneous network under study implemented in the EMTP-RV simulation environment. (Adapted from [41].)

cables. The series impedance and shunt admittance matrices for the line and cable are given by the following equations and have been calculated in correspondence with the line and cable switching frequency:

$$Z_{Line} = \begin{bmatrix} 1.10 + j15.32 & 1.00 + j5.80 & 1.00 + j4.64 \\ 1.00 + j5.80 & 1.09 + j15.33 & 1.00 + j5.80 \\ 1.00 + j4.64 & 1.00 + j5.80 & 1.10 + j15.32 \end{bmatrix} \frac{\Omega}{km}$$

(7.25)

$$Y_{Line} = \begin{bmatrix} 2 \times 10^{-4} + j67.53 & -j16.04 & -j7.91 \\ -j16.04 & 2 \times 10^{-4} + j70.12 & -j16.04 \\ -j7.91 & -j16.04 & 2 \times 10^{-4} + j67.53 \end{bmatrix}$$
$$\times 10^{-6} \frac{S}{km}$$

(7.26)

$$Z_{Cable} = \begin{bmatrix} 0.07 + j0.70 & 0.05 + j0.45 & 0.05 + j0.41 \\ 0.05 + j0.45 & 0.07 + j0.70 & 0.05 + j0.45 \\ 0.05 + j0.41 & 0.05 + j0.45 & 0.07 + j0.70 \\ 0.05 + j0.62 & 0.05 + j0.45 & 0.05 + j0.41 \\ 0.05 + j0.45 & 0.05 + j0.62 & 0.05 + j0.45 \\ 0.05 + j0.41 & 0.05 + j0.45 & 0.05 + j0.62 \end{bmatrix}$$

$$\begin{bmatrix} 0.05 + j0.62 & 0.05 + j0.45 & 0.05 + j0.41 \\ 0.05 + j0.45 & 0.05 + j0.62 & 0.05 + j0.45 \\ 0.05 + j0.41 & 0.05 + j0.45 & 0.05 + j.62 \\ 0.03 + j0.62 & 0.05 + j0.45 & 0.05 + j0.41 \\ 0.05 + j0.45 & 0.03 + j0.62 & 0.05 + j0.45 \\ 0.05 + j0.41 & 0.05 + j0.45 & 0.03 + j0.62 \end{bmatrix} \frac{\Omega}{km}$$

(7.27)

$Y_{Cable} =$

$$
\begin{bmatrix}
0.12 + j41.46 & 0 & 0 \\
0 & 0.12 + j41.46 & 0 \\
0 & 0 & 0.12 + j41.46 \\
-0.12 - j41.46 & 0 & 0 \\
0 & -0.12 - j41.46 & 0 \\
0 & 0 & -0.12 - j41.46 \\
\end{bmatrix}
$$

$$
\begin{bmatrix}
-0.12 - j41.46 & 0 & 0 \\
0 & -0.12 - j41.46 & 0 \\
0 & 0 & -0.12 - j41.46 \\
2.35 + j94.61 & 0 & 0 \\
0 & 2.35 + j94.61 & 0 \\
0 & 0 & 2.35 + j94.61 \\
\end{bmatrix} \times 10^{-6} \frac{S}{km}
$$

$$(7.28)$$

It can be observed that the simulated lines take into account the losses. The line left terminal and the cable right terminal are assumed to be terminated with power transformers represented, as discussed before, by high impedances, assumed, to a first approximation, equal to 100 kΩ. The supply of the line is provided by a three-phase AC voltage source placed at $x = 0$. All the fault transients were observed at the overhead line left terminal, shown as OP1, OP2, OP3 in Figure 7.20, corresponding to the three conductors of the line.

Two fault cases were considered to examine the performance of the proposed method for the case of inhomogeneous networks: (i) a three-phase-to-ground fault at 7 km away from the source with a 0 Ω fault impedance (solid) and (ii) a three-phase-to-ground fault at 5 km away from the source with a 100 Ω fault impedance (high-impedance fault). In agreement with the proposed procedure, the position of the GFL is moved along the overhead and cable lines assuming, for the fault impedance, a priori fixed values of 1, 10, and 100 Ω.

Figure 7.21 and Figure 7.22 show the energy of the current flowing through the GFL for solid and high-impedance faults, respectively. These figures show the calculated normalized FCSEs for three a priori guessed values of the fault resistance, namely 1, 10, and 100 Ω, as a result of the injection of the time-reversed voltage at the observation points (overhead line

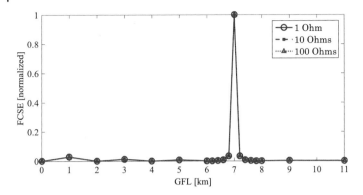

Figure 7.21 Normalized FCSE as a function of the GFL and for different guessed fault resistance values. The real fault location is at $x_f = 7$ km and real fault impedance is 0 Ω. (Adapted from [41].)

left terminal). In order to evaluate the accuracy of the proposed method, the position of the GFL is varied with a step of 200 m near to the real fault location.

It can be seen that the proposed method is effective in identifying the fault location in inhomogeneous networks, even when losses are present. The proposed method shows very good performances for high-impedance faults and also appears robust

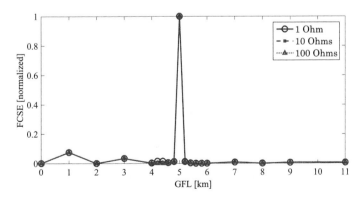

Figure 7.22 Normalized FCSE as a function of the GFL and for different guessed fault resistance values. The real fault location is at $x_f = 5$ km and real fault impedance is 100 Ω. (Adapted from [41].)

against the a priori assumed fault impedance. The accuracy of the method appears to be less than 200 m (assumed value for the separation between the GFL).

7.7.2 Series-Compensated Overhead Transmission Lines

In the last two decades, advances in power electronics have enabled sophisticated applications in power systems. In particular, applying series compensation in power systems can increase power transfer capability, improve transient stability, and damp power oscillations. However, series compensation introduces several technical challenges, specifically for the protection and fault location algorithms. For a series-compensated system, distance and fault-location estimation algorithms are significantly affected, leading to the malfunction of relays in different situations [56, 57].

Fault location in series-compensated transmission lines has a more crucial role since these types of lines are designed to link distant nodes among which a high amount of power is usually transferred. As summarized in [58], fault location methods for series-compensated transmission lines use either one- or multiple-end measurements and, in general, are based on post-fault impedance assessment.

In this section, the application of the proposed EMTR-based fault location method for series-compensated lines using single-end measurement is presented. To this end, an application example is considered by making reference to a three-conductor series-compensated transmission line. The line length is 200 km and the network is simulated in the EMTP-RV environment. The series compensation is done in the center of the transmission line to achieve a compensation degree of 50%. The relevant line parameters are the following:

- Positive and negative sequence impedance: $0.03293 + j0.3184$ Ω/km
- Positive and negative sequence capacitance: 0.01136 µF/km
- Zero sequence impedance: $0.2587 + j1.1740$ Ω/km
- Zero sequence capacitance: 0.00768 µF/km.

The line is assumed to be terminated at both ends on power transformers which, for signals characterized by high-frequency

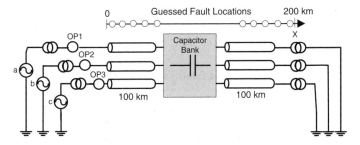

Figure 7.23 Schematic representation of the series-compensated three-conductor transmission line system implemented in the EMTP-RV. (Adapted from [59].)

spectrum content, can be replaced by high impedances (100 kΩ in this study).

The supply of the line is provided by a three-phase AC voltage source placed at $x = 0$. A schematic representation of the system is shown in Figure 7.23. In this figure, OP1, OP2, and OP3 are observation points corresponding to each conductor of the transmission line where voltage transients are recorded.

To examine the performance of the proposed method, two fault cases are considered: (i) a three-phase-to-ground fault at $x_f = 75$ km, (ii) a double phase-to-ground fault at $x_f = 35$ km. All these three fault cases are assumed to be solid faults.

By applying a similar procedure used in the previous application examples, the FCSE is calculated for each fault case. Figure 7.24a and b show the energy of the current flowing through the guessed fault points for the three-phase-to-ground fault and double-phase-to-ground fault, respectively. For each case, the energy values are normalized to the corresponding peak value. These figures illustrate the calculated normalized FCSEs for two a priori guessed values of the fault resistance, namely 1 Ω and 10 Ω.

It can be seen that the proposed method is remarkably effective in identifying the fault location for all three fault cases, even in the presence of the compensator (the maximum peak of the fault current energy is obtained at the real fault location). Additionally, the method appears robust against the a priori assumed fault impedance.

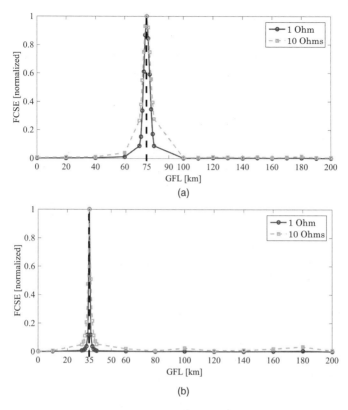

Figure 7.24 Normalized energy of the fault current as a function of GFLs and for different guessed fault resistance values (i.e., 1 and 10 Ω): (a) three-phase-to-ground fault at $x_f = 75$ km, (b) double-phase-to-ground fault at $x_f = 35$ km. (Adapted from [59].)

7.7.3   Radial Distribution Network: IEEE 34-Bus Test Distribution Feeder

In order to test the performance of the proposed fault location method in multibranch, multiterminal distribution networks, the IEEE 34-bus test feeder is considered. The model of this network is the same adopted in [26] where, for the sake of simplicity, the following assumptions have been made:

1) All transmission lines are considered to be characterized by configuration "ID #500" as reported in [60].

2) The loads are considered to be connected via secondary substation transformers located at line terminations.

Figure 7.25 shows the IEEE 34-bus test distribution network implemented in the EMTP-RV simulation environment.

This IEEE benchmark network is characterized by the presence of two transformers connected in series in two point of the system (V_REG_1 and V_REG_2 in Fig. 7.25). Their presence allows to control the amplitude of the voltage along the system during daily load variations. As already mentioned, with respect to high-frequency traveling waves, these transformers present a high input impedance. Therefore, the network can be partitioned in three zones each one equipped with a single observation point responsible to apply the proposed EMTR-based fault location technique. For this case study, only the first zone is considered, as shown in Figure 7.25. The observation point for this network is located at the secondary winding of the transformer and is shown in Figure 7.25.

Four different case studies are considered to examine the performance of the proposed method: (i) a three-phase-to-ground fault at Bus 808 with a 0 Ω fault impedance, (ii) a three-phase-to-ground fault at Bus 812 with a 100 Ω fault impedance, (iii) a single-phase-to-ground fault at Bus 810 with a 0 Ω fault impedance, and (iv) a single-phase-to-ground fault at Bus 806 with a 100 Ω fault impedance.

The recorded transient signals are time-reversed and, for each GFL, the current flowing through the fault resistance is calculated by simulating the network with back-injected time-reversed signals from the observation points. As in the previous cases, the normalized signal energy of this current is calculated for all GFLs with different guessed fault impedances (i.e., 1, 10, 100 Ω). Figure 7.26 shows the calculated FCSEs for (a) a three-phase-to-ground solid (0 Ω) fault at Bus 808 and (b) a three-phase-to-ground high-impedance fault (100 Ω) at Bus 812.

Figure 7.27 shows the calculated FCSE for (a) a single-phase-to-ground solid (0 Ω) fault at Bus 810 and (b) a single-phase-to-ground high-impedance fault (100 Ω) at Bus 806. From Figure 7.26 and Figure 7.27, it is possible to infer the remarkable performances of the fault location method for the case of realistic multibranch multiterminal lines. Additionally, the EMTR-based fault location method appears, also in this case, to be

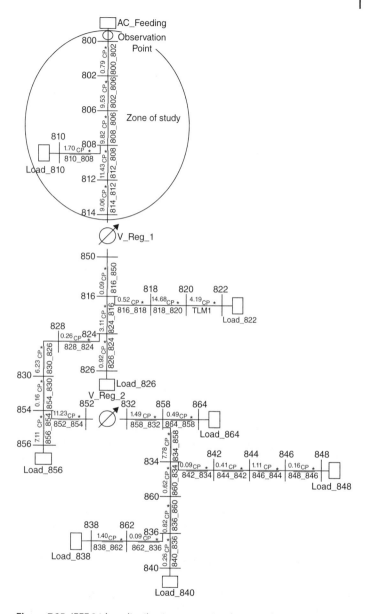

Figure 7.25 IEEE 34-bus distribution system implemented in EMTP-RV. (Adapted from [41].)

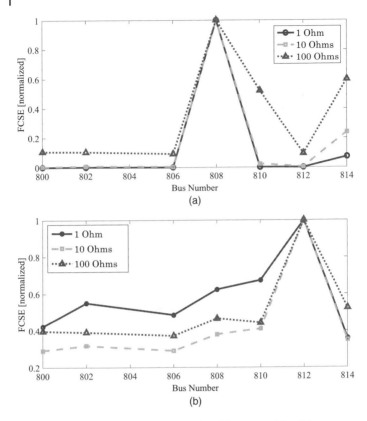

Figure 7.26 Normalized FCSE as a function of the GFL and for different guessed fault resistance values: (a) three-phase-to-ground solid fault (0 Ω) at Bus 808, (b) three-phase-to-ground high-impedance fault (100 Ω) at Bus 812. (Adapted from [41].)

robust against solid and high-impedance faults as well as against different fault types (phase-to-ground or three-phase ones).

7.7.4 Multiterminal HVDC Links

7.7.4.1 Background on Multiterminal HVDC Networks

Complex power networks topologies such as HVDC transmission systems enable massive integration of new types of generation units (i.e., renewable energy resources) available in remote

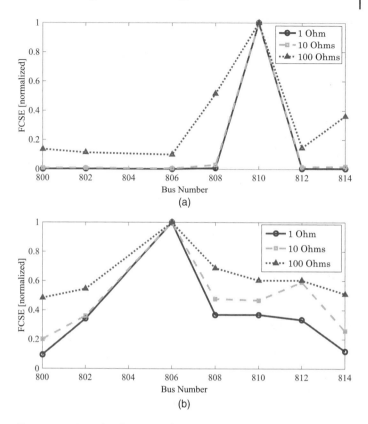

Figure 7.27 Normalized FCSE as a function of the GFL and for different guessed fault resistance values: (a) single-phase-to-ground solid fault (0 Ω) at Bus 810, (b) single-phase-to-ground high-impedance fault (0 Ω) at Bus 806. (Adapted from [41].)

locations (e.g., offshore wind farms). Traditional AC transmission systems are not the most suitable option to massively transfer the generated power for these distant generation units due to various technical and economical reasons [61]. Therefore, it is preferable to use DC transmission systems to interconnect large-scale distant wind farms.

An important challenge concerning large-scale offshore wind farms is the uncertainty, unpredictability, and variability of the generated power due the nature of the wind. One of the solutions to overcome this challenge is the interconnection of remote wind farms and other types of energy resources (e.g., hydro and solar) by means of multiterminal HVDC (MTDC) networks. As an example, the generated wind power in the Baltic Sea and North Sea can be balanced with the hydro pump-storage plants mainly located in Central Europe across the Alps (Southern Germany, Austria, Switzerland, Eastern France, and Northern Italy) and Norway [61, 62]. Furthermore, compared to point-to-point HVDC links, MTDC networks provide higher reliability of the power transfer and flexible power flow control between the stations and the AC grid [63].

Aside from the advantages provided by meshed MTDC networks, the protection and fault location problem represents the major challenge for the realization of these grids. Therefore, the fault location problem in MTDC networks requires more sophisticated processes (e.g., [61] and [64]). Indeed, in the protection schemes of traditional point-to-point HVDC links, current-regulating reactors limit the short-circuit currents and, additionally, the same protection scheme disconnects the system without the main selectivity requirements. However, for the case of MTDC networks, the rapid rise of the short-circuit current and the limited short-circuit current tolerance of the converters' anti-parallel diodes require the protection system to be characterized by an ultra-short time of intervention during which it has to discriminate the faulted line and then take the necessary countermeasures [65, 66]. In this respect, a few selective protection methods have been proposed in the literature, which are based on different techniques such as WT or differential protection (e.g., [65] to [70]).

As a consequence, MTDC protection systems need to be merged with fast fault location processes. This issue is particularly critical for the case of MTDC networks since the transmission lines are generally long and spread over seas, which limit the accessibility of the maintenance team. Furthermore, since the lines in such networks are considered to transfer bulk power over long distances, the loss of a line might cause overloading and congestion in other lines. A few fault location methods for the MTDC networks have been proposed in the literature (e.g., [67,68,71], and [72]). However, the investigations are in the early stages and more studies are necessary to be carried out.

The majority of the proposed fault location methods for HVDC networks are based on traveling waves, which provide accurate fault location estimates (e.g., [32] and [73]). However, compared to point-to-point HVDC links, the application of traveling wave-based fault location methods to MTDC networks could be challenging due to the following reasons:

- Due to multiple paths for the traveling waves in the MTDC networks, the fault location problem is more complicated compared to the case of two-terminal HVDC lines and can end up in multiple fault location identifications [26].
- Methods using multiterminal measurements provide, in general, more accurate and reliable results. However, they require multiple observation stations with time synchronization and fast communication links between them adding non-negligible complexity to the system.
- Sophisticated processing techniques employed by the conventional traveling wave-based fault location methods (e.g., WT, short-time Fourier transformation, etc.) might require considerable computational effort that does not necessarily match the limited time constraints associated with MTDC networks [27].

7.7.4.2 Fault Location in MTDC Networks Based on the EMTR

Application of the EMTR-based fault location method to the case of MTDC networks requires further adjustments. For the previous case studies, it has been assumed that the fault location system operates after the protection relays maneuver and that the time delay with respect to the fault occurrence of the protection system and the breakers opening are large enough so that the recorded transient signals are relatively damped out. For the case of MTDC networks, however, this assumption is no longer valid since the time scales of the HVDC protection systems intervention are much lower than those of the AC grids [74]. The fault identification and line disconnection time are mainly dependent on the protection system speed and the technology of the DC breakers, which are still evolving. Considering the typical time scales for the operation of HVDC protection system and the current interruption time required by the DC breakers, the total time to disconnect the faulty line is around 10 ms [65, 74]. As a consequence, the observation points located at the converter sides are not able to record the full time window of the fault-originated traveling waves.

After the opening of faulted line breakers, the system configuration and the boundary conditions associated with the traveling waves are changed. In this respect, it is worth noting that one of the main hypotheses in time-reversal theory is that *the system configuration should remain unchanged during the time-reversal process*. In addition, the breakers opening introduces additional surges that propagate in the system with additional traveling waves superimposed on the fault-originated ones. Therefore, the proposed EMTR-based fault location technique has to be suitably modified in order to overcome this problem. One straightforward approach to solve this problem is to limit the time window of the recorded transient signals up to the opening of the breakers, so that only the fault-originated traveling waves are considered.

As stated before, in order to consider only the fault-originated traveling waves, the time-reversal window is considered from the fault occurrence to the breaker opening. Therefore, only this part of the recorded voltage (which is shown in the expanded view – the waveform on the right in Figure 7.29) is considered for the EMTR fault location process. Then this signal is shifted in time and time-reversed.

The schematic representation of an MTDC network composed of five transmission lines and five converter stations is shown in Figure 7.28. As an example, let consider a pole-to-pole (p2p) fault at 10 km of line 1. The fault occurs at 10 ms and the line breakers disconnect the line at 20 ms. Figure 7.29 shows the recorded voltage for the positive node in station 5. In this figure, the red lines show the fault occurrence (10 ms) and the breakers disconnection time (20 ms).

Another challenge concerning MTDC networks is the existence of the multiple paths for the traveling waves and reflections from different boundaries. Therefore, it is appropriate to explore the possibility of using a limited time-reversal window in a medium with multiple reflective boundaries. This specific aspect is further discussed in the next section.

It has been shown that, when the time-reversal process is applied to a reflective medium, the boundaries of the medium

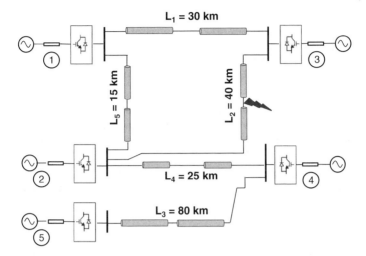

Figure 7.28 Schematic representation of the MTDC network under study.

can be considered as many virtual sources. For such boundary conditions, the information is confined in the system. Such peculiarity allows the process to be performed by using only one source in the back-propagation system [75, 76]. In fact, if the geometry of the medium shows *ergodic* and *mixing properties*, all information can be collected at only one observation point.

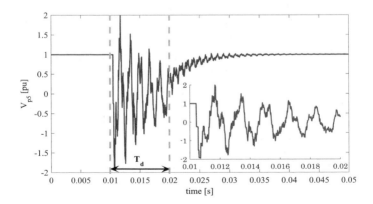

Figure 7.29 Recorded transient signals in station 5 for positive pole. The waveform on the right corresponds to the part of the signal from the fault occurrence to the breakers openings.

In other words, if the generated wave at a given point of the medium, after multiple reflections, passes every location of the medium, all the information about the source can be redirected towards a single observation point. Therefore, for systems having multiple reflective boundaries, like MTDC networks where the lines are terminated on power converter stations,[1] the process can be effectively applied by using only one observation point to record the transient signals and back-inject in the back-propagation model.

Concerning the impact of the time-reversal window on the time-reversal process performance for refocusing applications, Draeger and Fink [75] have shown that the duration of the back-injected signal (time-reversal window length) impacts the amplitude of the reconstructed wave. More specifically, the amplitude of the reconstructed waveform increases linearly with the time-reversal window duration (i.e., ΔT). On the other hand, the amplitude of the noise is a function of the square root of the time-reversal window (i.e., $\sqrt{\Delta T}$). Thus, the peak-to-noise ratio of the reconstructed signal increases with the square root of the time-reversal window [75]. As a consequence, the time reversal window impacts the sharpness of the reconstructed injected pulse. However, it is shown that the peak-to-noise ratio saturates for large time-reversal windows as multiple reflections cross over the observation point [75, 76]. For the case of acoustic cavities, the saturation time is reached after the Heisenberg time ($\tau_{Heisenberg}$), which is the minimum time duration to resolve each of the eigenmodes in the cavity [78]. In summary, in a closed reflecting medium, it is still possible to reconstruct the source by using a limited time-reversal window [75, 76, 78]. The duration of the time-reversal window is determined by the required refocusing quality. Larger time windows contain multiple reflections along the medium boundaries, which improve the time-reversal performance.

We take advantage of these two important peculiarities of the time-reversal process (single observation point and limited

---

1 Terminals of an HVDC grid where power converters are located are characterized by high values of input impedance for relatively high frequencies characterizing fault transients (e.g., [77]).

time window) to apply the proposed EMTR-based fault location method to the case of MTDC networks. Therefore, the recorded fault-originated transient signals at a given observation point are truncated at a maximum time corresponding to the faulty line disconnection time. Then the windowed transient signals are used to perform the fault location process.

To assess the performance of the proposed EMTR-based fault location method with a limited time-reversal window for the specific case of the MTDC networks reference is made to the network shown in Figure 7.28. A pole-to-pole (p2p) fault is considered at 20 km of the second line (the total line length is 40 km). As discussed earlier, for the case of MTDC networks, the time constraints for the protection systems is very extreme. Since the fault location procedure is performed after the fault clearance, we assume that the fault location system has the knowledge of the faulty line. Therefore, we narrow down the guessed fault locations (GFLs) investigation zone only to the faulty line.

By following the procedure presented earlier, and by considering only 10 ms of the fault-originated transient signals recorded at the observation point located at the 5th converter station, the fault current energies are calculated for different GFLs along the faulty line [79]. Figure 7.30 shows the fault current energies as a function of GFL for different time-reversal windows. Each curve

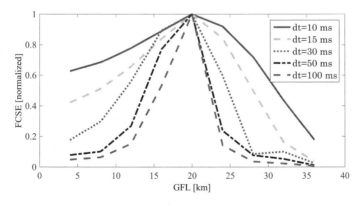

Figure 7.30 Impact of time-reversal window length on the accuracy of the proposed EMTR-based fault location method (the real fault location is in line 2 at 20 km).

is normalized to its own maximum value. It is observed that the increase in the time-reversal window duration results in a narrower peak around the localized fault point and provides better accuracy.

However, as can be seen by the presented results, a window duration of 10 ms, compatible with MTDC protection constraints, is still adequate to locate the fault. It is also worth noting that this time window is sufficiently long to record reflections coming from each end of the network.

In order to assess the impact of the noise on the performance of the EMTR-based fault location method, a case study is considered in which the fault-originated transient voltage signal is contaminated by the noise with a 20 dB signal-to-noise ratio (SNR). The fault is in line 2 at 20 km and the voltage transient signals are recorded at substation 5. Figure 7.31 shows the fault-originated transient signal recorded at station 5 for the positive pole without and with 20 dB noise.

Figure 7.32 shows the FCSE as a function of GFL for this case study. As can be observed, the performance of the fault location method is not influenced by the presence of the noise.

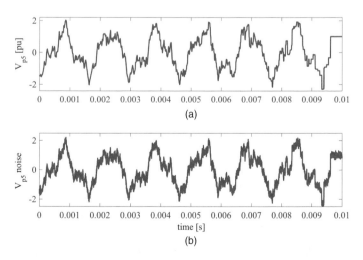

Figure 7.31 (a) The time-reversed voltage signal recorded at station 5 for positive pole; (b) the same signal by adding 20 dB noise.

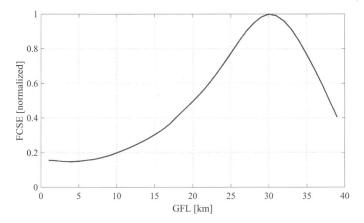

Figure 7.32 FCSE as a function of GFL. The fault is in line 2 at 30 km and the fault-originated transient signals recorded at substation 5 are contaminated with 20 dB SNR noise.

7.8 Conclusion

In this chapter, application of the electromagnetic time reversal (EMTR) to fault location in power grids has been presented. Compared with existing techniques, the EMTR-based method presents a number of advantages such as:

– Applicability to inhomogeneous and complex networks;
– Robustness against type and impedance of the fault, presence of noise and limited observation time window;
– Use of a single observation point.

The EMTR-based fault location method is carried out in three steps: (1) measurement of the fault-originated electromagnetic transients in a single observation point, (2) simulation of the back-injection of the time-reversed measured fault signal for different guessed fault locations and using the network model, and (3) determination of the fault location by computing, in the simulated network, the point characterized by the largest fault current energy signal associated with the back-injected time-reversed fault transients.

The method is validated by means of reduced scale experiments considering two topologies, namely one single transmission line and a T-shape network. In both cases, the proposed EMTR-based approach was able to correctly identify the location of the fault. It is worth observing that these experiments were performed in the presence of the cable losses and measurement noise.

Several validation examples have been performed by making reference to different types of power networks including (i) an inhomogeneous network composed of mixed overhead lines and coaxial cables, (ii) a radial distribution network, (iii) a series-compensated transmission line, and (iv) a multiterminal HVDC network. The resulting fault location accuracy and robustness against uncertainties (e.g., fault impedance, fault type, etc.) have been tested.

References

1 M. M. Saha, J. J. Izykowski, and E. Rosolowski, *Fault Location on Power Networks*, vol. 25. Springer Science & Business Media, 2009.

2 "IEEE Guide for Determining Fault Location on AC Transmission and Distribution Lines," 2005.

3 T. W. Stringfield, D. J. Marihart, and R. F. Stevens, "Fault location methods for overhead lines," *Transactions of the American Institute of Electrical Engeers, Part III: Power Apparatus Systems*, vol. 76, no. 3, pp. 518–529, April 1957.

4 T. Takagi, Y. Yamakoshi, M. Yamaura, R. Kondow, and T. Matsushima, "Development of a new type fault locator using the one-terminal voltage and current data," *IEEE Transactions on Power Apparatus Systems*, vol. PAS-101, no. 8, pp. 2892–2898, August 1982.

5 L. Eriksson, M. M. Saha, and G. D. Rockefeller, "An accurate fault locator with compensation for apparent reactance in the fault resistance resulting from remote-end infeed," *IEEE Power Engineering Review*, vol. PER-5, no. 2, pp. 44–44, February 1985.

6 C. E. M. Pereira and L. C. Zanetta, "Fault location in transmission lines using one-terminal postfault voltage data,"

IEEE Transactions on Power Delivery, vol. 19, no. 2, pp. 570–575, April 2004.

7 T. Kawady and J. Stenzel, "A practical fault location approach for double circuit transmission lines using single end data," *IEEE Transations on Power Delivery*, vol. 18, no. 4, pp. 1166–1173, October 2003.

8 M. Kezunovic and B. Perunicic, "Automated transmission line fault analysis using synchronized sampling at two ends," in *Proceedings of Power Industry Computer Applications Conference*, pp. 407–413, 1995.

9 A. A. Girgis, D. G. Hart, and W. L. Peterson, "A new fault location technique for two- and three-terminal lines," *IEEE Transactions on Power Delivery*, vol. 7, no. 1, pp. 98–107, 1992.

10 D. Novosel, D. G. Hart, E. Udren, and J. Garitty, "Unsynchronized two-terminal fault location estimation," *IEEE Transactions on Power Delivery*, vol. 11, no. 1, pp. 130–138, 1996.

11 A. L. Dalcastagne, S. N. Filho, H. H. Zurn, and R. Seara, "An iterative two-terminal fault-location method based on unsynchronized phasors," *IEEE Transactions on Power Delivery*, vol. 23, no. 4, pp. 2318–2329, October 2008.

12 J. Izykowski, E. Rosolowski, P. Balcerek, M. Fulczyk, and M. M. Saha, "Accurate noniterative fault-location algorithm utilizing two-end unsynchronized measurements," *IEEE Transactions on Power Delivery*, vol. 26, no. 2, pp. 547–555, April 2011.

13 T. Nagasawa, M. Abe, N. Otsuzuki, T. Emura, Y. Jikihara, and M. Takeuchi, "Development of a new fault location algorithm for multi-terminal two parallel transmission lines," *IEEE Transactions on Power Delivery*, vol. 7, no. 3, pp. 1516–1532, July 1992.

14 G. Manassero, E. C. Senger, R. M. Nakagomi, E. L. Pellini, and E. C. N. Rodrigues, "Fault-location system for multiterminal transmission lines," *IEEE Transactions on Power Delivery*, vol. 25, no. 3, pp. 1418–1426, July 2010.

15 T. Funabashi, H. Otoguro, Y. Mizuma, L. Dube, and A. Ametani, "Digital fault location for parallel double-circuit multi-terminal transmission lines," *IEEE Transactions on Power Delivery*, vol. 15, no. 2, pp. 531–537, April 2000.

16 S. M. Brahma, "Fault location scheme for a multi-terminal transmission line using synchronized voltage measurements,"

IEEE Transactions on Power Delivery, vol. 20, no. 2, pp. 1325–1331, April 2005.

17 C.-W. Liu, K.-P. Lien, C.-S. Chen, and J.-A. Jiang, "A universal fault location technique for N-terminal transmission lines," *IEEE Transactions on Power Delivery*, vol. 23, no. 3, pp. 1366–1373, July 2008.

18 J. Izykowski, R. Molag, E. Rosolowski, and M. M. Saha, "Accurate location of faults on power transmission lines with use of two-end unsynchronized measurements," *IEEE Transactions on Power Delivery*, vol. 21, no. 2, pp. 627–633, April 2006.

19 Y. Liao and N. Kang, "Fault-location algorithms without utilizing line parameters based on the distributed parameter line model," *IEEE Transactions on Power Delivery*, vol. 24, no. 2, pp. 579–584, April 2009.

20 Y. G. Paithankar and M. T. Sant, "A new algorithm for relaying and fault location based on autocorrelation of travelling waves," *Electrical Power Syststems Research*, vol. 8, no. 2, pp. 179–185, March 1985.

21 P. McLaren and S. Rajendra, "Travelling-wave techniques applied to the protection of Teed circuits: multi-phase/multi-circuit sytstem," *IEEE Transactions on Power Apparatus Systems*, vol. PAS-104, no. 12, pp. 3551–3557, December 1985.

22 A. O. Ibe and B. J. Cory, "A travelling wave-based fault locator for two- and three-terminal networks," *IEEE Transactions on Power Delivery*, vol. 1, no. 2, pp. 283–288, 1986.

23 A. M. Ranjbar, A. R. Shirani, and A. F. Fathi, "A new approach for fault location problem on power lines," *IEEE Transactions on Power Delivery*, vol. 7, no. 1, pp. 146–151, 1992.

24 F. H. Magnago and A. Abur, "Fault location using wavelets," *IEEE Transactions on Power Delivery*, vol. 13, no. 4, pp. 1475–1480, 1998.

25 G. B. Ancell and N. C. Pahalawaththa, "Maximum likelihood estimation of fault location on transmission lines using travelling waves," *IEEE Transactions on Power Delivery*, vol. 9, no. 2, pp. 680–689, 1994.

26 A. Borghetti, M. Bosetti, M. Di Silvestro, C. A. Nucci, and M. Paolone, "Continuous-wavelet transform for fault location in distribution power networks: definition of mother wavelets inferred from fault originated transients," *IEEE Transactions on Power Systems*, vol. 23, no. 2, pp. 380–388, 2008.

27 Y. Zhang, N. Tai, and B. Xu, "Fault analysis and traveling-wave protection scheme for bipolar HVDC lines," *IEEE Transactions on Power Delivery*, vol. 27, no. 3, pp. 1583–1591, 2012.

28 E. H. Shehab-Eldin and P. G. McLaren, "Travelling wave distance protection-problem areas and solutions," *IEEE Transactions on Power Delivery*, vol. 3, no. 3, pp. 894–902, July 1988.

29 D. J. Spoor and J. G. Zhu, "Improved single-ended traveling-wave fault-location algorithm based on experience with conventional substation transducers," *IEEE Transactions on Power Delivery*, vol. 21, no. 3, pp. 1714–1720, July 2006.

30 M. Ando, E. Schweitzer, and R. Baker, "Development and field-data evaluation of single-end fault locator for two-terminal HVDV transmission lines – Part 2: Algorithm and evaluation," *IEEE Transactions on Power Apparatus Systems*, vol. PAS-104, no. 12, pp. 3531–3537, December 1985.

31 F. V. Lopes, K. M. Silva, F. B. Costa, W. L. A. Neves, and D. Fernandes, "Real-time traveling-wave-based fault location using two-terminal unsynchronized data," *IEEE Transactions on Power Delivery*, vol. 30, no. 3, pp. 1067–1076, June 2015.

32 M. B. Dewe, S. Sankar, and J. Arrillaga, "The application of satellite time references to HVDC fault location," *IEEE Transactions on Power Delivery*, vol. 8, no. 3, pp. 1295–1302, 1993.

33 C.-S. Yu, "An unsynchronized measurements correction method for two-terminal fault-location problems," *IEEE Transactions on Power Delivery*, vol. 25, no. 3, pp. 1325–1333, July 2010.

34 A. Borghetti, M. Bosetti, C. A. Nucci, M. Paolone, and A. Abur, "Integrated use of time-frequency wavelet decompositions for fault location in distribution networks: theory and experimental validation," *IEEE Transactions on Power Delivery*, vol. 25, no. 4, pp. 3139–3146, 2010.

35 A. Borghetti, S. Corsi, C. A. A. Nucci, M. Paolone, L. Peretto, and R. Tinarelli, "On the sse of continuous-wavelet transform for fault location in distribution power systems," *International Journal of Electrical Power Energy Systems*, vol. 28, no. 9, pp. 608–617, November 2006.

36 M. Vetterli and C. Herley, "Wavelets and filter banks: theory and design," *IEEE Transactions on Signal Processes*, vol. 40, no. 9, pp. 2207–2232, 1992.

37 F.-C. Lu, Y. Chien, J. P. Liu, J. T. Lin, P. H. S. Yu, and R. R. T. Kuo, "An expert system for locating distribution system faults," *IEEE Transactions on Power Delivery*, vol. 6, no. 1, pp. 366–372, 1991.

38 K. K. Kuan and K. Warwick, "Real-time expert system for fault location on high voltage underground distribution cables," *IEEE Proceedings C, General Transactions on Distribution*, vol. 139, no. 3, pp. 235–240, 1992.

39 J.-C. Maun, "Artificial neural network approach to single-ended fault locator for transmission lines," in *Proceedings of the 20th International Conference on Power Industry Computer Applications*, pp. 125–131, 1997.

40 J. Gracia, A. J. Mazon, and I. Zamora, "Best ANN structures for fault location in single- and double-circuit transmission lines," *IEEE Transactions on Power Delivery*, vol. 20, no. 4, pp. 2389–2395, October 2005.

41 R. Razzaghi, G. Lugrin, H. M. Manesh, C. Romero, M. Paolone, and F. Rachidi, "An efficient method based on the electromagnetic time reversal to locate faults in power networks," *IEEE Transactions on Power Delivery*, vol. 28, no. 3, pp. 1663–1673, July 2013.

42 H. M. Manesh, G. Lugrin, R. Razzaghi, C. Romero, M. Paolone, and F. Rachidi, "A new method to locate faults in power networks based on electromagnetic time reversal," in *2012 IEEE 13th International Workshop on Signal Processing Advances in Wireless Communications (SPAWC)*, pp. 469–474, 2012.

43 A. Greenwood, *Electrical Transients in Power Systems*. New York: John Wiley & Sons, Inc., 1991.

44 A. Borghetti, M. Bosetti, C. A. Nucci, M. Paolone, and A. Abur, "Fault location in active distribution networks by means of the continuous-wavelet analysis of fault-originated high frequency transients," in *Proceedings of the Cigré General Session*, 2010.

45 M. Fink, C. Prada, F. Wu, and D. Cassereau, "Self focusing in inhomogeneous media with time reversal acoustic mirrors," in *Proceedings, IEEE Ultrasonics Symposium*, pp. 681–686, 1989.

46 M. Fink, "Time reversal of ultrasonic fields. I. Basic principles," *IEEE Transactions on Ultrasonic Ferroelectric Frequency Control*, vol. 39, no. 5, pp. 555–566, January 1992.

47 F. Rachidi, C. A. Nucci, M. Ianoz, and C. Mazzetti, "Influence of a lossy ground on lightning-induced voltages on overhead lines," *IEEE Transactions on Electromagnetic Compatibility*, vol. 38, no. 3, pp. 250–264, 1996.

48 F. Rachidi, "A review of field-to-transmission line coupling models with special emphasis on lightning-induced voltages on overhead lines," *IEEE Transactions on Electromagnetic Compatibility*, vol. 54, no. 4, pp. 898–911, August 2012.

49 M. Fink, "Time reversal of ultrasonic fields. I. Basic principles," *IEEE Transactions on Ultrasonic Ferroelectric Frequency Control*, vol. 39, no. 5, pp. 555–566, January 1992.

50 L. Gaspard, R. Razzaghi, F. Rachidi, and M. Paolone, "Electromagnetic time reversal applied to fault detection: the issue of losses," in *Joint IEEE International Symposium on Electromagnetic Compatibility and EMC Europe*, 2015.

51 C. A. Nucci and F. Rachidi, "Interaction of electromagnetic fields generated by lightning with overhead electrical networks," in *The Lightning Flash*, V. Cooray, ed., pp. 559–610. London: IET, 2004.

52 G. Lugrin, R. Razzaghi, F. Rachidi, and M. Paolone, "Electromagnetic time reversal applied to fault detection: the issue of losses," in *2015 IEEE International Symposium on Electromagnetic Compatibility (EMC)*, pp. 209–212, 2015.

53 H. Dommel, "Digital computer solution of electromagnetic transients in single-and multiphase networks," *IEEE Transactions on Power Appararatus Systems*, vol. PAS-88, no. 4, pp. 388–399, April 1969.

54 J. Mahseredjian, S. Lefebvre, and X.-D. Do, "A new method for time-domain modelling of nonlinear circuits in large linearnetworks," in *Proceedings of the 11th Power Systems Computation Conference (PSCC)*, 1993.

55 J. Mahseredjian, S. Dennetière, L. Dubé, B. Khodabakhchian, and L. Gérin-Lajoie, "On a new approach for the simulation of transients in power systems," *Electrical Power Systems Research*, vol. 77, no. 11, pp. 1514–1520, September 2007.

56 C.-S. Yu, C.-W. Liu, S.-L. Yu, and J.-A. Jiang, "A new PMU-based fault location algorithm for series compensated lines," *IEEE Transactions on Power Delivery*, vol. 17, no. 1, pp. 33–46, 2002.

57 P. Jena and A. K. Pradhan, "A positive-sequence directional relaying algorithm for series-compensated line," *IEEE Transactions on Power Delivery*, vol. 25, no. 4, pp. 2288–2298, October 2010.

58 J. Izykowski, E. Rosolowski, P. Balcerek, M. Fulczyk, and M. M. Saha, "Fault location on double-circuit series-compensated lines using two-end unsynchronized measurements," *IEEE Transactions on Power Delivery*, vol. 26, no. 4, pp. 2072–2080, October 2011.

59 R. Razzaghi, G. Lugrin, M. Paolone, and F. Rachidi, "On the use of electromagnetic time reversal to locate faults in series-compensated transmission lines," in *2013 IEEE Grenoble Conference*, pp. 1–5, 2013.

60 W. H. Kersting, "Radial distribution test feeders," in *2001 IEEE Power Engineering Society Winter Meeting. Conference Proceedings (Cat. No. 01CH37194)*, vol. 2, pp. 908–912, 2001.

61 D. Van Hertem and M. Ghandhari, "Multi-terminal VSC HVDC for the European supergrid: obstacles," *Renewable and Sustainable Energy Review*, vol. 14, no. 9, pp. 3156–3163, 2010.

62 P. Bresesti, W. L. Kling, and R. Vailati, "Transmission expansion issues for offshore wind farms integration in Europe," in *2008 IEEE/PES Transmission and Distribution Conference and Exposition*, pp. 1–7, 2008.

63 C. M. Franck, "HVDC circuit breakers: a review identifying future research needs," *IEEE Transactions on Power Delivery*, vol. 26, no. 2, pp. 998–1007, April 2011.

64 J. Yang, J. E. Fletcher, and J. O'Reilly, "Multiterminal DC wind farm collection grid internal fault analysis and protection design," *IEEE Transactions on Power Delivery*, vol. 25, no. 4, pp. 2308–2318, 2010.

65 J. Descloux, B. Raison, and J. B. Curis, "Protection strategy for undersea MTDC grids," in *2013 IEEE Grenoble Conference on PowerTech, POWERTECH 2013*, 2013.

66 N. Ahmed, A. Haider, D. Van Hertem, L. Zhang, and H.-P. Nee, "Prospects and challenges of future HVDC SuperGrids with modular multilevel converters," in *Proceedings of the 2011 -14th European Conference on Power Electronic Applications (EPE 2011)*, pp. 1–10, 2011.

67 L. Tang and B.-T. Ooi, "Locating and ssolating DC faults in multi-terminal DC systems," *IEEE Transactions on Power Delivery*, vol. 22, no. 3, pp. 1877–1884, July 2007.

68 K. De Kerf, K. Srivastava, M. Reza, D. Bekaert, S. Cole, D. Van Hertem, and R. Belmans, "Wavelet-based protection strategy for DC faults in multi-terminal VSC HVDC systems," *IET Generation and Transmission Distribution*, vol. 5, no. January, p. 496, 2011.

69 L. T. L. Tang and B.-T. Ooi, "Protection of VSC-multi-terminal HVDC against DC faults," in *2002 IEEE 33rd Annual IEEE Power Electronic Special Conference Proceedings (Cat. No. 02CH37289)*, vol. 2, pp. 719–724, 2002.

70 A. E. B. Abu-Elanien, A. A. Elserougi, A. S. Abdel-Khalik, A. M. Massoud, and S. Ahmed, "A differential protection technique for multi-terminal HVDC," *Electrical Power Systems Research*, vol. 130, pp. 78–88, January 2016.

71 O. M. K. K. Nanayakkara, A. D. Rajapakse, and R. Wachal, "Traveling-wave-based line fault location in star-connected multiterminal HVDC systems," *IEEE Transactions on Power Delivery*, vol. 27, no. 4, pp. 2286–2294, October 2012.

72 J. Yang, J. E. Fletcher, and J. O'Reilly, "Short-circuit and ground fault analyses and location in VSC-based DC network cables," *IEEE Transactions on Industrial Electronics*, vol. 59, no. 10, pp. 3827–3837, October 2012.

73 J. Suonan, S. Gao, G. Song, Z. Jiao, and X. Kang, "A novel fault-location method for HVDC transmission lines," *IEEE Transactions on Power Delivery*, vol. 25, no. 2, pp. 1203–1209, April 2010.

74 W. Leterme and D. Van Hertem, "Classification of fault clearing strategies for HVDC grids," in *Cigre*, 2015.

75 C. Draeger and M. Fink, "One-channel time reversal of elastic waves in a chaotic 2D-silicon cavity," *Physical Review Letters*, vol. 79, no. 3, pp. 407–410, July 1997.

76 C. Draeger, J.-C. Aime, and M. Fink, "One-channel time-reversal in chaotic cavities: experimental results," *Journal of the Acoustical Society of America*, vol. 105, no. 2, p. 618, February 1999.

77 Y. Zhang, N. Tai, and B. Xu, "Fault analysis and traveling-wave protection scheme for bipolar HVDC lines," *IEEE Transactions on Power Delivery*, vol. 27, no. 3, pp. 1583–1591, July 2012.

78 M. Fink, "Time-reversal acoustics in complex environments," *Geophysics,* vol. 71, no. 4, p. SI151, 2006.

79 R. Razzaghi, M. Paolone, F. Rachidi, J. Descloux, B. Raison, and N. Retière, "Fault location in multi-terminal HVDC networks based on electromagnetic time reversal with limited time reversal window," in *18th Power Systems Computation Conference (PSCC),* 2014.

Index

Electromagnetic Time Reversal: Application to Electromagnetic Compatibility and Power Systems, First Edition. Edited by Farhad Rachidi, Marcos Rubinstein and Mario Paolone.
© 2017 John Wiley & Sons, Ltd. Published 2017 by John Wiley & Sons, Ltd.
Companion Website: www.wiley.com/go/rachidi55